MW00955719

# Chinese Tactics

August 2021

United States Government
US Army

# Contents

---

DISTRIBUTION RESTRICTION: Approved for public release, distribution is unlimited.

## PART TWO    PEOPLE'S LIBERATION ARMY ACTIONS

# Figures

# Tables

# Preface

ATP 7-100.3 describes Chinese tactics for use in Army training, professional education, and leader development. This document is part of the ATP 7-100 series that addresses a nation-state's military doctrine with a focus on army ground forces and tactical operations in offense, defense, and related mission sets. Other foundational topics include task organization, capabilities, and limitations related to military mission and support functions. ATP 7-100.3 serves as a foundation for understanding how Chinese ground forces think and act in tactical operations. This publication presents multiple examples of functional tactics in dynamic operational environment conditions. The tactics in this ATP are descriptive, and provide an orientation to tactics gathered from Chinese doctrine, translated literature, and observations from recent historical events.

The principal audience for ATP 7-100.3 is all members of the profession of arms. Commanders and staffs of Army headquarters serving as joint task force or multinational headquarters should also refer to applicable joint or multinational doctrine concerning the range of military operations and joint or multinational forces. Trainers and educators throughout the Army will also use this publication.

Commanders, staffs, and subordinates ensure that their decisions and actions comply with applicable United States, international, and in some cases host-nation laws and regulations. Commanders at all levels ensure that their Soldiers operate in accordance with the law of war and the rules of engagement. (See FM 6-27.)

To compare and contrast information in this ATP with other Army doctrine, the reader must first understand the fundamentals of land operations in FM 3-0 and the Army's supporting ADPs and ATPs that describe military operations and the application of combat power. Joint and multinational application will require comparison to and contrast with relevant joint and multinational doctrine.

ATP 7-100.3 uses joint terms where applicable. Selected joint and Army terms and definitions appear in both the glossary and the text. For definitions shown in the text, the term is italicized and the number of the proponent publication follows the definition. This publication is not the proponent for any Army terms.

Chinese concepts and publications are mentioned throughout this publication. The former are underlined upon either first or second appearance, and the latter appear in italics. When a U.S. term—either joint or Army—has the same name as a Chinese concept and the U.S. term is being referenced, it appears in italics.

ATP 7-100.3 applies to the Active Army, Army National Guard/Army National Guard of the United States, and United States Army Reserve unless otherwise stated.

The proponent of ATP 7-100.3 is the United States Army Combined Arms Center. The preparing agency is the Combined Arms Doctrine Directorate, United States Army Combined Arms Center. The lead agency tasked with developing this ATP is the United States Army Training and Doctrine Command G-2, Analysis and Control Element, Operational Environment and Threat Analysis Directorate. Send comments and recommendations on DA Form 2028 (Recommended Changes to Publications and Blank Forms) to Commander, United States Army Combined Arms Center and Fort Leavenworth, ATTN: ATZL-MCD (ATP 7-100.3), 300 McPherson Avenue, Fort Leavenworth, KS 66027-2337; by email to usarmy.leavenworth.mccoe.mbx.cadd-org-mailbox@mail.mil; or submit an electronic DA Form 2028.

# Introduction

ATP 7-100.3, as part of the U.S. Army 7-100 series and in support of AR 350-2, addresses the tactics, organization, and activities of the People's Republic of China's (PRC's) military, the People's Liberation Army (PLA). The PRC is a near-peer competitor to the United States, with significant political and strategic interests worldwide, though the PRC's primary geographic focus is Eastern and Southeastern Asia and the Western Pacific. The PRC has developed significant capabilities in all domains: land, air, maritime, space, and cyberspace. Capabilities in specific operational environments include a number of different unit and system combinations able to challenge the combat power of U.S. and allied forces. The PLA represents a significant competitor to U.S. and allied forces, especially in the Western Pacific. This publication provides a doctrinal description of PLA tactics and organizations in order to facilitate robust, realistic, relevant, and challenging conditions in which to train U.S. and allied forces.

## PART ONE—PEOPLE'S LIBERATION ARMY FORCES

**Chapter 1** describes the strategic environment, operational environments, China's political objectives, and its associated strategy. It then delves into China's military forces: its philosophical underpinnings, principles, approach to conflict, operational framework, and system warfare.

**Chapter 2** describes the PLA's force structure for its land combat components, an overview of its theater-level capabilities, and a discussion of tactical-level units such as brigades and battalions.

**Chapter 3** provides an overview of PLA joint capabilities that can affect tactical operations, including air forces, naval forces, rocket and missile forces, and specialized support forces.

**Chapter 4** describes the PLA's primary framework for operations: system warfare, planning processes, control measures, command posts, and combat groups.

**Chapter 5** describes how the PLA approaches information operations, with an emphasis on actions at tactical echelons.

## PART TWO—PEOPLE'S LIBERATION ARMY ACTIONS

**Chapter 6** describes the PLA approach to reconnaissance, counterreconnaissance, and security in support of tactical operations at tactical echelons.

**Chapter 7** describes PLA offensive operations at tactical echelons and how the system warfare approach is applied in these actions.

**Chapter 8** describes PLA defensive operations at tactical echelons and how the system warfare approach is applied in in these actions.

**Chapter 9** describes the PLA approach to stability operations at tactical echelons and how the system warfare approach is applied during stability actions.

## APPENDIXES

**Appendix A** describes the capabilities and limitations of PLA maneuver forces: infantry, armor, marine, and airborne.

**Appendix B** describes PLA indirect fire capabilities, including mortars, tube artillery, and rocket artillery in support of group army and tactical operations; this includes an overview of integrated fires systems.

**Appendix C** describes PLA tactical air defense forces controlled by its land forces, including an overview of integrated air defense systems up to the group army echelon.

**Appendix D** describes how the PLA aviation assets—primarily rotary-wing manned aircraft and unmanned aircraft—support tactical operations.

**Appendix E** describes the use of engineers and chemical defense forces in support of tactical operations.

**Appendix F** describes the PLA's communications and network architecture at the tactical level.

**Appendix G** describes special operations forces both at tactical and national/strategic echelons.

# UNITS OF MEASURE

Units of measure in ATP 7-100.3 are metric. The introductory table provides conversion data from one measurement system to another for units used in this publication.

**Introductory Table. Unit conversions**

| Type | Metric | English |
|------|--------|---------|
| Distance | 1 meter (m) | 3.28084 feet |
| Distance | 1 kilometer (km) | 0.62137 miles |
| Weapon Bore Size | 1 millimeter (mm) | 0.03937 inches |
| Weapon Bore Size | 1 millimeter (mm) | 0.03937 caliber |
| Weight | 1 kilogram (kg) | 2.2046 pounds |
| Weight | 1 metric ton (t) | 1.1023 (U.S./short) tons |

**PART ONE**

# People's Liberation Army Forces

Part one describes the intellectual and philosophical framework of the People's Liberation Army (PLA), describes Chinese political objectives and strategy, provides an overview of PLA joint capabilities, and describes the PLA's system warfare construct and unit structure in detail. The information in this section provides the theoretical structure necessary to fully understand the PLA's tactical approach, and also provides detailed information necessary to create realistic variables as a part of building a Chinese threat model.

---

**Chapter 1**

# People's Liberation Army Fundamentals

The People's Liberation Army (PLA) continues more than two millennia of Chinese military tradition. China lays cultural claim to many of the world's most famous works of military strategy and philosophy—most prominently, *The Art of War* by Sun Tzu—and these works are still highly influential throughout the PLA. The Chinese take military philosophy, politics, and theory very seriously: the PLA is considered the vanguard of the Chinese Communist Revolution, and its philosophical underpinnings are important to both its culture and its approach to warfighting at all levels. The PLA also has a deep commitment to Communist and Maoist philosophy.

## THE STRATEGIC ENVIRONMENT

1-1.   China's view of the strategic environment mirrors that of the United States in many ways. There are, however, key differences in both analysis of the strategic environment and the application of this analysis that underpin important differences in perspective between the two countries. Both the People's Republic of China—commonly referred to as China—and the United States assess the key elements of the strategic environment discussed in paragraphs 1-2 through 1-9.

1-2.   **U.S. obligations to allies and partners in the Western Pacific will continue in perpetuity.** Key U.S. allies include Japan, South Korea, Australia, New Zealand, and the Philippines. The United States will likewise maintain a strong strategic interest in the Western Pacific.

1-3.   **The United States will maintain strong, though unofficial, support to the Republic of China**— also known as Taiwan. The former will employ a policy of deliberate ambiguity to deter potential Chinese aggression against the island while maintaining a position respectful of Chinese policy toward the same.

1-4.   **China will continue to seek improved relations with Russia and India, with Russia likely proving a more amenable partner.** China views improving these two relationships—particularly with Russia—as very important both politically and economically. Border tensions with India complicate bilateral relations and are often perceived by India as aggressive, requiring strong responses and adjustments to its defense posture.

1-5.   Increasing competition for limited resources and the effects of global climate change will fuel conflict among both state and non-state actors. As populations increase, providing adequate quantities of basic necessities such as water, energy, food, and medical care becomes increasingly difficult. In addition, competition for human capital and raw materials for industry will increase as the international economy expands. Conditions such as natural disasters, extreme weather events, and their second- and third-order effects will have considerable impacts across the globe and will significantly influence political and military strategies worldwide. The global commons—the earth's unowned natural resources, such as the oceans, the atmosphere, and space—will be increasingly important to the global economy and will thus likely become contested. The PLA characterizes many tasks related to these conditions under the heading of nontraditional security measures or military operations other than war.

1-6.   **As the global economy and disparate societies become increasingly interconnected, friction points continue to emerge.** State and non-state actors are already using widely proliferated but immature technologies—such as social networking— worldwide to influence populations, politics, and policy. Actors will continue to leverage emerging connections to manipulate public opinion and influence leaders; the digital world will become an increasingly important front to contest. As national economies continue to intertwine themselves, competition for jobs, natural and manmade resources, and new or emerging markets will continue to intensify.

1-7.   **Technological advances will continue to enhance the lethality of capabilities across all domains.** Soldiers at all echelons on the future battlefield will face unprecedented dangers as both new and proven technologies are integrated into warfighting. For example, precision munitions will continue to proliferate and become increasingly affordable as technology costs drop. At the same time, limited defense budgets coupled with the high expense and relative rarity of modern weapons systems—particularly aircraft, surface ships, and missiles of all types—will heavily influence future major combat scenarios.

1-8.   **China faces both nontraditional and hybrid threats.** Such threats include criminal organizations, terrorist organizations, and fringe or rogue states. These threats effectively exploit complex terrain, such as dense urban and cyber environments, and are not necessarily constrained by international law or protocols regulating conflict.

---

*Note.* The chapters and appendixes of this publication address topics from the Chinese perspective. As such, the terms *friendly* refers to Chinese units or the units of Chinese allies. *Enemy* refers to units opposing China or its military. This may be a belligerent force or element within China, or an external actor. Parties are *neutral* regarding China. Threat uses the standard dictionary definition as opposed to that of U.S. doctrine. An opponent may be against either the U.S. or China, with context determining the correct interpretation.

---

1-9.   **Four strategic trends will influence future Chinese operations.** These trends are population growth, urbanization, population growth along major bodies of water, and human connectedness and interrelations.

---

*Note.* Chinese concepts and doctrinal phrases contained in this document are used in accordance with PLA definitions as provided by the translation of *Army Combined Tactics under Conditions of Informationization* and *Infantry Unit Tactics*. Some of these concepts and phrases may mirror the names of U.S. Department of Defense (DOD) and U.S. Army terms. In such cases, the U.S. term is being referenced only if the name appears in *italics*.

---

# CHINESE POLITICAL OBJECTIVES

1-10. A useful analysis of the tactics of the PLA must be underpinned by an understanding of Chinese political and strategic priorities and an understanding of the complex relationship between the PLA and the Communist Party of China (CPC). China's history is deeply intertwined with its military—more so than any Western nation and possibly more so than any other country, except North Korea. The PLA was the vanguard of the Chinese Communist Revolution; its history, from the Chinese perspective, is one of glorious struggle over imperialist and capitalist oppressors. In contrast to Western militaries, the PLA is deeply politicized. It retains significant ties to Maoist and Marxist-Leninist political thought and has generations-deep connections with the CPC. While Western militaries are generally apolitical and are divorced from political parties, the PLA is officially the armed wing of the CPC.

> **Note.** This is an explanation of naming and acronym conventions. The proper name for China's military is the People's Liberation Army, or PLA. This organizational structure is comparable to the U.S. Department of Defense (DOD); it does not refer exclusively to ground forces, as does the U.S. Army. The PLA's land-based service is the People's Liberation Army Army (PLAA), the naval branch is the People's Liberation Army Navy (PLAN), and the aerial branch is called the People's Liberation Army Air Force (PLAAF). The PLA's strategic conventional and nuclear ballistic and cruise missile service is the People's Liberation Army Rocket Force (PLARF). The People's Liberation Army Strategic Support Force (PLASSF) is a new service-level organization that incorporates national-level cyber warfare, electronic warfare (EW), space, and other operational support elements under a single headquarters, and the People's Liberation Army Joint Logistics Support Force (PLAJLSF) is a new service-level sustainment organization.

1-11. Despite China's gradual economic liberalization and movement toward a capitalist, free-market economy, the PLA ostensibly remains generally steadfast in its support of Marxist-Leninist and Maoist philosophy. Though the PLA long ago abandoned more extreme Maoist experiments such as "rankless" force structures and eliminating rank insignia, the idea of People's War—an armed struggle of a population against a militarily superior adversary—still colors PLA thinking. This concept eventually took the form of People's War under Modern Conditions, an adaptation of Chinese Communist populist warfare that accounts for modern military capabilities. This in turn evolved to People's War in Conditions of Informationization in 2015. This evolution suggests that the PLA now sees itself as having acquired sufficient information capabilities to successfully employ them in a limited or regional military capacity. Marxist-Leninist and Maoist thinking still influence PLA operations all the way down to the tactical level, where a mix of autocratic statism and communal leadership are practiced even within small organizations.

1-12. Chinese national political objectives can be broken into two basic categories: security and development. Security objectives include the protection of the CPC as China's ruling party, the protection of Chinese sovereignty, protection of borders, internal security, and nuclear deterrence. Development objectives include the protection of Chinese economic interests at home and abroad, ensuring freedom of navigation for Chinese goods, procuring important commodities such as energy and raw materials, and establishing new export markets for Chinese goods. As China has transitioned from a closed, unstable, post-revolutionary internal political environment to a more open and stable one, strategic priorities have gradually shifted from security to development.

1-13. Today, with the primacy of the CPC virtually assured and few internal security threats, the People's Armed Police (PAP) has taken over much of the internal security mission, while the PLA focuses primarily on development-related objectives. The coordination of military modernization with economic development is a basic tenet of Chinese national strategy, resulting in significant annual defense budget increases for the past two decades. This principle is now manifest in its Military-Civil Fusion program, which seeks greater cooperation between military and civilian elements in achieving shared objectives. An important objective of Military-Civil Fusion is leveraging civilian assets and capabilities as a cost-efficient method of managing limited military resources. Military-Civil Fusion can be thought of as an extension of People's War, seeking to make best use of civilian assets and natural resources to underpin the growth and modernization of the PLA.

1-14. The ongoing Chinese movement toward development objectives requires a significant enhancement of PLA capabilities. Throughout much of its history, including its active conflicts in Korea (1950-53) and Vietnam (1979), the PLA had little to no expeditionary capability, minimal mechanization, low-technology systems, and a severe lack of military professionalism. Objectives that support development, however, generally require meaningful force projection capabilities: a blue-water navy, a modern air force, long-range targeting and strike capabilities, enhanced long-range air and sea strategic lift capabilities, and a well-trained, well-equipped, professional ground force. Due to one of the largest and longest economic growth cycles in history, China was able to fund the reform and development of the PLA, modernizing it over the last three decades from a force of poorly armed and poorly trained conscripts into a viable modern military.

## THE NEW CHINESE GRAND STRATEGY

1-15. China desires status as a world power, likely using Imperial Great Britain or post-World War II United States as a model. This end state helps to achieve many smaller goals important to the Chinese: global influence, economic development, internal security, and CPC primacy. China views this final objective as incremental: it seeks to be a "prosperous society" by 2035, with the CPC still remaining the dominant political entity in China. This objective likely implies China being a regional hegemon in the Western Pacific, with a robust middle class and fully modernized military. The country desires to transition to a "leading world power" by the year 2049 (the centennial anniversary of the country's founding), complete with a "world-class military." These objectives are not nebulous, nor are they fanciful thinking—they are written into the CPC Constitution. Of note, despite the nature of these objectives, the Chinese government does not describe itself as seeking either regional or global hegemony.

1-16. Throughout Chinese history, beginning in the third century B.C. through the Second Sino-Japanese War (1937-45), incursions, invasions, and occupations along China's vast land border and coastline were constant. This fostered a deep-set national sense of sovereignty and border sanctity among the Chinese. Even in the Imperial Age (pre-1912), Chinese leaders sought to offset external threats through a combination of political savvy and military strength. China was historically the dominant economic and military power in East Asia, and Chinese leaders used this power to try and ensure domestic security and tranquility.

1-17. Much of this economic and military power was lost during the colonial and world-war periods, a time the Chinese refer to as a "century of humiliation." China's transition to a Communist political and economic system was a long and enormously destructive process that further marginalized its once-dominant power. It was not until the late 1970s that the country's resurgence really began. During this recent period of enormous economic growth and political change, China again began to seek a position as a world power. Though the global environment has changed through the years, the Chinese approach to security and economic growth has not.

1-18. China today seeks what it views as a restoration of its position as a global power through what are described by the CPC as peaceful and relatively unprovocative means. Bilateral and multilateral relationships are viewed as optimal, where all parties to a matter see benefit from an arrangement. China seeks a positive relationship with the West—especially the United States—underpinned by a massive exchange of trade and economic interdependence. China also seeks to modernize its military—not through aggressive, short-term arms buildup, but through long-term investment and development. It also seeks courteous—if not cordial—relations with its neighbors in East Asia. In short, China wishes to be a good neighbor and a good global citizen.

1-19. This approach has thus far yielded mixed results. Territorial disputes across the Western Pacific and East Asia, aggressive cyber activities, uncooperative diplomacy, questionable trade practices, and a horrendous human rights record all undermine the Chinese goal of being seen as a benevolent superpower. Activities that China views as fundamentally defensive—such as the establishment of artificial island airbases in the South China Sea—are seen as aggressive and provocative by virtually every other party in the region. Chinese cyber activities, seen by the nation as simple surveillance and deterrence activities, have caused several high-profile international incidents.

## CHINESE STRATEGIC OBJECTIVES

1-20. China's strategic objectives are informed by its political objectives and generally support the broad political goals of Western Pacific dominance by 2035 and becoming a leading world power by 2049. They are also informed by certain elements of Chinese culture and history: the importance of status and honor, the desire for peace through power, and the belief in Chinese Communism, among other aspects. Chinese strategic objectives are to—

- Maintain internal security and stability.
- Secure and protect land borders and coastlines.
- Maintain regional stability.
- Maintain freedom of navigation.
- Resolve maritime territorial disputes.
- Establish positive conditions for potential hostilities.

# MAINTAIN INTERNAL SECURITY AND STABILITY

1-21. Foremost in maintaining internal security and stability is ensuring the position of the CPC as the dominant political entity in China. PLA headquarters at provincial, county, and city levels are military components of the corresponding local government and are responsible for recruitment, demobilization, and other support to local governments. While maintaining domestic stability is the primary mission of the civilian police force and PAP, in extreme situations the PLA may be required to assist these forces in internal security operations.

## SECURE AND PROTECT LAND BORDERS AND COASTLINES

1-22. China's long historic struggle with border security manifests itself today in a vast array of border security measures. Of particular note is the border with North Korea: while ostensibly friendly with the North Koreans, China is deeply concerned with the possible military and humanitarian crisis that might erupt in the event that the North Korean regime falls or war breaks out on the Korean Peninsula. In addition, China's shared border with India continues to be a source of friction and conflict.

## MAINTAIN REGIONAL STABILITY

1-23. One of the main benefits China sees to regional dominance is the maintenance of peace and order in the Western Pacific and East Asia. In addition to the complex political and diplomatic challenges, this requires the PLA to be capable of conducting shaping and deterrence operations throughout the region, able to deploy and defeat regional threats as required.

## MAINTAIN FREEDOM OF NAVIGATION

1-24. China's economy is largely dependent on exports, most of which travel via the world's oceans. Maintaining safe and free passage through the global commons is a necessary component of Chinese political stability and economic development. Particular focus areas include the main shipping channels of the Western Pacific, such as the Strait of Malacca and the South Indian Ocean.

## RESOLVE MARITIME TERRITORIAL DISPUTES

1-25. China views several land masses and their surrounding territorial waters in the South China Sea—and elsewhere in the Pacific—as strategically important. These include, but are not limited to, the Senkaku Islands, the Spratly Islands, and the Paracel Islands. Their importance is derived not only from their proximity to important global shipping lanes, but also due to their potential usefulness as military bases—particularly for naval and coast guard ships, fixed-wing aircraft, antisubmarine warfare capabilities, and land-based antiaircraft and antiship missile systems.

## ESTABLISH POSITIVE CONDITIONS FOR POTENTIAL HOSTILITIES

1-26. China views conflict along a continuum, ranging from steady-state deterrence operations in peacetime through major combat operations. By meticulously—and sometimes covertly—conducting peacetime military and intelligence operations against potential opponents, China seeks to place its military in an advantageous position should active hostilities break out. Such operations include political and diplomatic efforts, offensive and defensive cyber actions, information operations, and covert intelligence operations. Central to this objective is the expansion of overseas basing for the PLA, enabling force projection outside of Chinese borders and giving Chinese leadership greater flexibility in choosing how and where to employ military force.

> *Note.* The concept of information operations will be used in this document consistent with the Chinese expression *xinxi zuozhan* (信息作战). This is similar to the U.S. DOD term *information operations*, though it is not as inclusive. Chinese information operations include information warfare, concealment, deception (general efforts to mislead an opponent), and trickery (specific plans targeted at a particular opponent). The Chinese expression for information warfare, *xinxi zhan* (信息战), refers to direct, specific offensive and defensive actions, such as EW and cyber warfare, and is not analogous to the U.S. concept of information warfare that pertains to an opposing force.

## OPERATIONAL ENVIRONMENTS

1-27. U.S. military analysis of an operational environment, including a composite environment created for training, professional education, and leader development purposes, focuses on eight interrelated operational variables: political, military, economic, social, information, infrastructure, physical environment, and time—collectively referred to as PMESII-PT. The following is a list of PMESII-PT conditions and trends found in China. It is not comprehensive, but it suggests a number of possible factors that U.S. exercise planners might use when constructing scenarios:

- Political:
    - The fractious relationship between China and Taiwan.
    - Maritime disputes in the South and East China Seas.
    - Complex sociopolitical interactions between China and North Korea.
    - Land border disputes and frictions.
    - Friction between the CPC and the growing quasi-capitalist ultra-wealthy class.
    - Expansion of Chinese influence in emerging markets, particularly in Africa and Southwest Asia.
    - Human rights violations, particularly against internal political opposition and minorities, and the harassment and mistreatment of journalists.
    - Separatism in Western China.
    - Major anticorruption efforts at every level of government.
- Military:
    - Growing use of high-technology weapons systems such as fifth-generation and low-observable aircraft, ballistic and cruise missiles, precision munitions, and networked warfare.
    - Ongoing top-to-bottom reform of the PLA, including widespread professionalization.
    - Expansion of standoff precision munitions and other antiaccess capabilities.
    - Establishment of hardened military facilities on islands, both natural and manmade, throughout the Western Pacific.
    - Gradual expansion of PLA expeditionary capabilities, particularly throughout the Western Pacific and Indian Oceans.
- Economic:
    - Continued liberalization of formerly hardline Marxist-Leninist and Maoist economic policy.
    - Threats to critical shipping lanes and overland trade routes linking China with its export markets.
    - Export and trade restrictions on raw materials, particularly on rare but critical metals.
    - Major anticorruption efforts at every level of government.
    - Complex economic relationship with North Korea.

- Social:
  - Increased internal frictions with minority groups, particularly in and around distressed border regions.
  - Continued adoption of Western cultural products, especially by younger generations.
  - Social resistance to heavy-handed governmental approach to internal security.
  - Increased frictions between quasi-capitalist Chinese oligarchs and more-traditional CPC supporters due to failure to quell corruption.
- Information:
  - Extensive use of cyber activities—official, unofficial, and third party—to influence conditions domestically and abroad.
  - Ongoing active People's Republic of China (PRC) cyber activities attempting to extract sensitive or classified information from foreign networks, particularly those in defense industries.
  - Sophisticated information operations campaigns to influence both global and regional politics.
- Infrastructure:
  - Continued development of domestic infrastructure with a focus on the export economy.
  - Heavy investment in overseas infrastructure, particularly in emerging markets for Chinese export goods.
- Physical Environment:
  - Ongoing effects of global climate change reducing availability of arable land and threatening low-lying coastal areas.
  - Increasing frequency and severity of tropical storms, which affect military and economic development in the Western Pacific.
  - Shortsighted environmental policies creating public health crises due to air and water pollution and rapid depletion of shared international resources, such as fisheries.
- Time:
  - The Chinese have historically taken a much longer view of time than the United States.
  - The PLA has deliberately chosen to adopt a more Western view of time as part of its ongoing military reforms.

# THE PEOPLE'S LIBERATION ARMY—OVERVIEW

1-28. The PLA's basic warfighting philosophy is that of <u>active defense</u>: a fundamentally defensive political and strategic stance, enabled—when required—by operational and tactical offense. For over two thousand years, China has been surrounded by enemies, adversaries, and other competitors. Invasion, occupation, raids, and other incursions into Chinese territory were commonplace. The PLA views protecting Chinese sovereignty and security as a sacred duty. China traditionally viewed military resistance as an affair for the entire population: mass resistance, guerrilla warfare, and winning a war of attrition. This understanding has evolved in the modern age to where a PLA enabled by technology, well-trained personnel, and a whole-of-government, defense-in-depth approach deters conflict before it ever happens and protects China and the CPC from foreign aggression and internal tumult. The basic concept of active defense informs every level of PLA operations and acquisitions. Figure 1-1 provides a graphic depiction of the policy and theoretical underpinnings of the active defense philosophy.

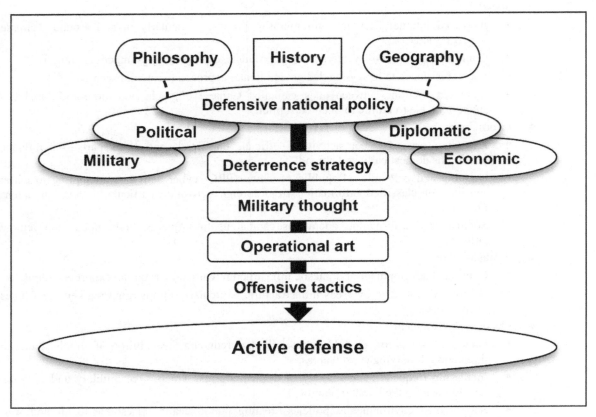

**Figure 1-1. Active defense**

1-29. Presently, the PLA is in the midst of a period of comprehensive reform. Central to this is the evolution of the "big army"—the dominance of the PLAA at the expense of other services—not only throughout the PLA, but also throughout Chinese politics and society. Reducing manpower and equipment levels and employing a quality-over-quantity approach is central to this effort, as is the expansion of joint integration and an expanded emphasis on domains other than just the ground. In addition, PLAA units are now nationally deployable—able to operate anywhere within Chinese borders—rather than being strongly tied to a specific region. This is intended to reduce manpower requirements and the influence of the PLA in local politics, but it requires a top-to-bottom reorganization of PLA training, doctrine, and acquisition.

1-30. Two historical figures feature prominently in the philosophy and strategy of the PLA: Sun Tzu and Mao Zedong. The PLA views these two strategists as equally important, with People's War being viewed as a modern addendum to *The Art of War*. Sun Tzu and Mao together created the framework from which the PLA's modern-day strategy and tactics are derived.

1-31. Influenced largely by *The Art of War*, the PLA—and China as a whole—look to create advantages along a vast competition continuum, ranging from diplomacy and covert operations in peacetime to major combat operations. The PLA views all available government means as a kind of defense in depth. Defense of China begins with skilled diplomacy and prudent political measures at home and abroad. State actors and agencies set conditions for military success if and when military action becomes necessary. The PLA decisively defeats opponents militarily if peaceful measures fail. The ideal outcome in every scenario is to "win without fighting"—as a philosophy taken directly from Sun Tzu. Military-Civil Fusion is an important method of incorporating civilian support to achieve military objectives in both peacetime and war.

1-32. PLA doctrine describes three basic ways to use military power: warfighting, military deterrence, and military operations other than war. Developing warfighting capabilities is the PLA's core task; prevention of conflict is the PLA's most important mission. The PLA's deterrence mission creates numerous subordinate objectives designed to demonstrate Chinese capabilities and will, all viewed as critical to preventing conflict both domestically and abroad. Though China states unequivocally that it will not initiate conflict, practices only self-defense, and will respond militarily only if attacked, its basic concept of deterrence has been broadened significantly in recent years. Active defense now includes far more politically and strategically offensive activities. Examples include, but are not limited to—

- Expanded participation in global peacekeeping operations.
- Expanded air and naval presence in the Western Pacific in an offshore defensive strategy.
- More rapid and thorough responses to non-war crises.
- Expanded operations in the cyber and space domains.
- Defense of the primacy and stability of the CPC.

1-33. PLA views on modern war allow for limited preemptive actions. People's War initially mandated few to no pre-emptive military activities, but PLA views on modern war concede that an unmitigated modern military operation may win a war before a proper response can be organized. As such, certain allowances for offensive activity under the umbrella of active defense have been made. According to PLA doctrine, the "first shot" in the area of politics and strategy must be differentiated from the first shot on the plane of tactics. If any country or organization violates the other country's sovereignty and territorial integrity, the other side will have the right to "fire the first shot" on the plane of tactics.

1-34. People's War encompasses all three levels of warfare. At the strategic level, People's Armies use their superior manpower, lesser logistic requirements, and greater political will to defeat opponents in defensive wars of attrition. At the operational level, a combination of mobility, deception, and willpower allow formations to rapidly maneuver and close with the enemy in order to engage it in close combat, offsetting limitations in firepower, technology, and training. At the tactical level, units aggressively maneuver in order to create massive advantages in combat power at key times and locations, while simultaneously preventing enemy response through effective deception operations. Infiltration techniques allow small units to close with and defeat the enemy in detail. At all levels, People's Army units rely on superior willpower, cohesion, deception, and manpower to overcome technologically advanced opponents.

1-35. People's War was initially developed as Mao's interpretation of Marxism in military conflict: a class struggle between an oppressed people and a bourgeois or imperialist professional military. The Second Sino-Japanese War and the Chinese Civil War (1945-49) strengthened this understanding: both conflicts involved a People's Army conducting mobile or guerrilla operations, on its own soil, against better-equipped and trained opponents. Both conflicts resulted in Chinese victories, but they were enormously costly, both economically and in lives lost.

1-36. People's War concept and doctrine were significantly altered beginning in the 1980s. At the end of the Mao era in the late 1970s, PLA strategists began to reassess the nature of People's War. Mao had taken Marxist-Leninist ideals to extremes in the military, adopting a military without hierarchy or rank. This approach proved ineffective, and reforms were needed to make the PLA into a modern force capable of deterring conflict and defending Chinese strategic and political interests. This evolved into People's War under Modern Conditions, a modification that accounted for the increasing lethality and complexity of the modern battlefield. Joint operations—along with mechanization and artillery—were emphasized. The concept of deterrence expanded to include not only fighting defensive wars within Chinese territory, but also the idea that a strong regional and maritime security environment might prevent fighting in Chinese territory in the first place.

1-37. The modern iteration of People's War is People's War in Conditions of Informationization. This theory incorporates thinking and strategy informed by numerous geopolitical and technological developments over the past 25 years. First and foremost is the emerging information environment: the PLA recognizes the importance of the digital battlefield, and it has emerged as a world leader in integrating cyber warfare into its operational construct. Figure 1-2 on page 1-10 displays the timeline of the evolution of People's War concepts.

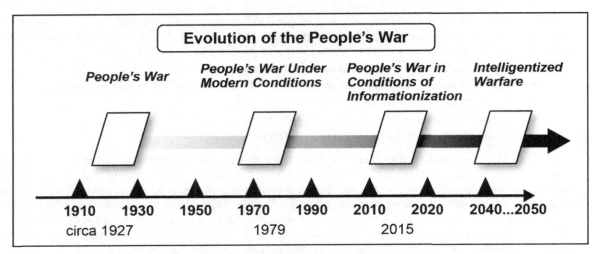

**Figure 1-2. People's War**

1-38. Second, the PLA further expanded its understanding of active defense. After watching nearly six months of force buildup on the Iraqi border during OPERATIONS DESERT SHIELD and DESERT STORM, the PLA concluded that the most efficient way to defend Chinese territory was to prevent potential opponents from building combat power in close proximity to Chinese borders. This in turn spurred the development and redevelopment of capabilities intended to defeat or interdict the buildup of combat power at extended ranges and across wide geographic areas. It also compelled development of capabilities designed to deny use of wide geographic areas to potential foes. Years later, in analyzing this approach, U.S. strategists denoted it *antiaccess/area denial*. This phrase was—and still is—an accurate depiction of the PLA approach to active defense in the Western Pacific.

1-39. Third, People's War in Conditions of Informationization encompasses the tacit admission that China may have to become involved in local or regional conflicts in order to maintain the regional or international economic order, protect the CPC, or otherwise support Chinese economic or political interests. This ostensibly contradicted Mao's initial philosophy of People's War, which is for the most part strictly a domestic defense-oriented military approach. This appreciation may represent fundamental change for the PLA, and it requires a vast increase in expeditionary and overseas sustainment capabilities. These requirements, in turn, demand significant increases in the PLA's military professionalization, equipment quality, and air and naval capabilities.

1-40. Local War is the PLA expression for a regional conventional conflict, likely close to Chinese territory. The PLA views this as the most likely situation calling for military force, and it has developed capabilities designed to excel in this environment. This conclusion makes sense, as every conflict the PLA has fought in its history was local, and all but the Chinese Civil War were limited. People's War in Conditions of Informationization, when applied to the Local War concept, is frequently referred to as Winning Informationized Local Wars.

1-41. Chinese military theory may be moving away from traditional Marxist-Leninist and Maoist theory as PLA leaders acknowledge the importance of technology, professionalization, and military hierarchy. The PLA, however, still identifies as staunchly Communist and views its modern theories as an evolution of People's War, not as a revision or repudiation. People's War in Conditions of Informationization is said to be a modern adaptation of *The Art of War* and People's War. It retains the concept of active defense as its centerpiece, and it considers deception and political willpower to be the most important—and most uniquely Chinese—elements of a successful military campaign.

1-42. The PLA anticipates Informationized Warfare evolving into Intelligentized Warfare in the relatively near future. Intelligentized Warfare incorporates numerous emerging technologies—including decentralized computing, data analytics, quantum computing, artificial intelligence, and unmanned or robotic systems—into the PLA's conceptual framework. Intelligentized Warfare seeks to increase the pace of future combat by effectively fusing information and streamlining decision-making, even in ambiguous or highly dynamic operating environments. Intelligentized Warfare also amplifies the nascent concepts embodied by the Military-Civil Fusion effort: many of the subsystems that create the backbone of an Intelligentized PLA are researched and developed initially in the civilian realm. Careful alignment of military and civilian efforts enables the synchronization of efforts and streamlines the fielding process for the PLA. Despite its focus on technology, Intelligentized Warfare remains informed by the People's War concepts: many Intelligentized Warfare initiatives are clearly shaped by the original People's War principles.

# PLA PRINCIPLES—PEOPLE'S WAR

1-43. PLA principles were originally written by Mao Zedong during the Long March of 1934—35 and revised during the Japanese occupation of China, beginning in 1937. These principles still serve as the basis for People's War theories, though they have been modernized periodically along with the rest of Chinese military thought. There are numerous different interpretations and translations of these principles, varying widely based on when and where they were written and translated. However, the versions are all similar, generally reflecting Communist political sensibilities, a focus on mobility and deception, and a strong understanding of basic military theory. The key themes of People's War are—

- Eliminate isolated pockets of the enemy before concentrating to fight larger forces.
- Capture small villages and towns before capturing large urban areas.
- Eliminate the enemy's fighting capacity; do not focus on territory.
- Fight no battle unprepared; develop strategy based on the worst conditions.
- Concentrate forces to achieve an overwhelming advantage in numbers, then defeat the enemy in detail.
- Choose the first battle carefully.
- Unify the command and coordinate.
- Combine mobile war, positional war, and guerrilla war.
- Employ forces and tactics flexibly.
- Fight in one's own way, and let the enemy fight in its.

## ELIMINATE ISOLATED POCKETS OF THE ENEMY BEFORE CONCENTRATING TO FIGHT LARGER FORCES

1-44. This principle is similar to Western maneuver warfare theories developed during the World Wars. It calls for rapid maneuver and decisive attacks against weak points before finally seeking a decisive engagement against an unbalanced enemy.

## CAPTURE SMALL VILLAGES AND TOWNS BEFORE CAPTURING LARGE URBAN AREAS

1-45. This somewhat unique Communist Chinese principle was formed from the success of early PLA operations against Japanese and Republican Chinese forces. Operating in rural areas made it difficult for early PLA opponents to detect and fix PLA formations. In expeditionary warfare, this principle likely implies targeting less-challenging defensive positions first, then moving on to more well-defended and critical positions once initial gains are consolidated.

## ELIMINATE THE ENEMY'S FIGHTING CAPACITY; DO NOT FOCUS ON TERRITORY

1-46. This principle is similar to the principle in paragraph 1-45 about towns and cities. It recognizes that, for most military campaigns, the center of gravity is the enemy force, not territory. Chinese forces may temporarily—or even deliberately—cede territory in order to enable decisive action against the enemy's forces.

## FIGHT NO BATTLE UNPREPARED; DEVELOP STRATEGY BASED ON THE WORST CONDITIONS

1-47. This principle is similar to the general planning process of many armed forces. Commanders ensure that troops are as prepared as possible for any battle, and they conduct planning based on the enemy's most dangerous course of action, rather than its most probable.

## CONCENTRATE FORCES TO ACHIEVE AN OVERWHELMING ADVANTAGE IN NUMBERS, THEN DEFEAT THE ENEMY IN DETAIL

1-48. This is another principle similar to Western military theories, though it specifies the importance of numerical advantage. This principle is most often seen in PLAA tactics: tactical-level units seek to use a combination of maneuver and deception to achieve their desired numerical superiority and allow engagement and defeat of the enemy in detail. Traditionally, the PLA sought to fight using close combat techniques, taking advantage of night operations to enable infiltration whenever possible. Technological advances changed this approach, incorporating longer-range weapons and nonlethal effects to strike the enemy at greater distances, decreasing the effect of the technological gaps the PLA faces with regard to night vision and electro-optical capabilities.

## CHOOSE THE FIRST BATTLE CAREFULLY

1-49. This principle acknowledges the importance of gaining and maintaining the initiative; in this case, through a victorious first engagement. This principle extends beyond simple military battles; it also includes choosing the first decisive political or economic engagement carefully in order to set conditions for victory in a future decisive engagement. Along with "fight no battle unprepared," this principle suggests an element of caution and prudence in determining when and if to go to war.

## UNIFY THE COMMAND AND COORDINATE

1-50. As a basic and timeless military principle, PLA organization and doctrine regarding unified command differs somewhat from Western militaries. While the latter imbue command authority with a single individual—the commander—the PLA considers unity of command to be as much a political issue as an operational principle. Units feature both military and political leadership, and creating political unity of purpose is considered a critical component of People's War.

## COMBINE MOBILE WAR, POSITIONAL WAR, AND GUERRILLA WAR

1-51. This principle outlines the three types of warfare proposed by Mao, and it resembles the more modern theory of hybrid warfare. Mobile units use a combination of maneuver and deception to achieve surprise in decisive actions; static, defensive units fix opponents and defend or secure key areas from attack and guerrilla forces conduct irregular campaigns either behind enemy lines or within enemy formations. All forms are carefully coordinated to ensure unity of purpose and situational understanding.

## EMPLOY FORCES AND TACTICS FLEXIBLY

1-52. Similar to U.S. Army principles and doctrine, this principle acknowledges the importance of flexibility in any military situation. Historically, the PLA was notoriously inflexible, with subordinates often failing to employ their own judgment due to fear of reprisal or lack of professional competence. The PLA's new focus on building a professional noncommissioned officer corps and a decentralized leadership philosophy reflect its understanding of shortfalls in this area. Many PLA exercises currently stress tactical innovation on the part of commanders at all levels.

## FIGHT IN ONE'S OWN WAY, AND LET THE ENEMY FIGHT IN ITS

1-53. Perhaps the most famous of Mao's principles, this principle reinforces China's views about the uniqueness of its approach to military operations. It stresses the importance of knowing both one's own capabilities and those of the opponent.

# CHINA'S APPROACH TO CONFLICT

1-54. China considers three aspects in the country's view of conflict. They are Comprehensive National Power, deception, and the Three Warfares.

## COMPREHENSIVE NATIONAL POWER

1-55. China describes warfare as only one waypoint along a continuum. Military power is only one component of Comprehensive National Power (CNP), the Chinese method of ranking countries based on an assessment of all types of state influence. Hard power includes military capability and capacity, defense industry capability, intelligence capability, and related diplomatic actions such as threats and coercion. Soft power includes such things as economic power, peaceful diplomatic efforts, foreign development, global image, and international prestige. These two types of power combine to make up CNP. China views CNP as a vital measure of its global status. Ultimately, all forms of conflict—be they military, diplomatic, or other—must enhance China's CNP. CNP is viewed as a whole-of-government effort; the PLA is simply the military wing of CPC, so it makes sense to view military action as fundamentally political. The primary mission of the PLA is strategic deterrence, as illustrated in figure 1-3.

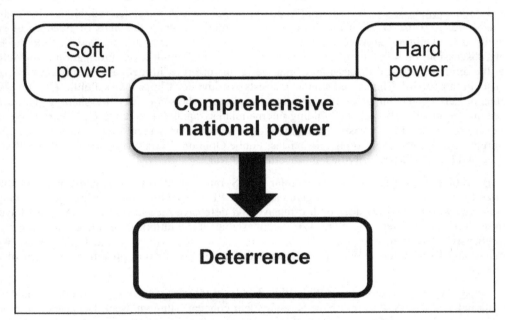

Figure 1-3. Comprehensive national power

## DECEPTION

1-56. Deception plays a critical role in every part of the Chinese approach to conflict. The Chinese emphasis on deception can be traced to Sun Tzu, who believed that it was the basis for all warfare. PLA views on this topic differ considerably from those of most Western militaries. Instead of being a peripheral enabler, deception operations are seen as integral to every operation at all levels of war. Where U.S. Army operational planning uses the concept of a *course of action*— a scheme developed to accomplish a mission (JP 5-0)—PLA planners use stratagems. Rather than describing friendly operations, stratagems describe the enemy's mindset, focusing on how to achieve the desired perceptions by the opponent, and then prescribing ways to exploit this perception. Rather than focusing on defeating the opponent in direct conflict—as most Western militaries do—stratagems consider deception, trickery, and other indirect, perception-based efforts to be the most important elements of an operation. Deception is a fundamental aspect of the Chinese way of war, and applications of deception are considered a high priority.

## THREE WARFARES

1-57. China's strategic approach to conflict employs <u>Three Warfares</u> designed to support and reinforce the PLA's traditional military operations. These Three Warfares are—

- Public Opinion Warfare.
- Psychological Warfare.
- Legal Warfare.

Though these approaches are called warfares, these strategies—in Western thinking—fall somewhere between modern concepts such as information operations and historical concepts such as military operations other than war or effects-based operations. Despite the names, they are universally nonlethal: they do not involve direct combat operations. Instead, they are designed to pursue what Sun Tzu considered generalship in its highest form—victory without battle. If a battle must be fought, the Three Warfares are designed to unbalance, deceive, and coerce opponents in order to influence their perceptions. In a major change from the past, when political officers were mainly involved in rear area personnel functions, the Three Warfares make political officers and soldiers into nonlethal warfighters who provide essential support to combat units.

1-58. <u>Public Opinion Warfare</u> is referred to as *huayuquan*, which translates roughly as "the right to speak and be heard." To the Western mind this implies something along the lines of freedom of speech. Its meaning to the Chinese, however, is substantially different: it refers to the power to set the terms of a debate, discussion, or negotiation. In other words, it is China's high-level information campaign designed to set the terms of political discussion. China views this effort as influential not only on PLA operations, but also in support of Chinese economic interests worldwide. China views Public Opinion Warfare as capable of seizing the initiative in a conflict before any shots are fired by shaping public discourse, influencing political positions, and building international sympathy. Public Opinion Warfare operations are seen every day in the PLA's vast media system of newspapers, magazines, television, and internet sources that target both domestic and foreign audiences. Public Opinion Warfare supports the PLA's Psychological Warfare and Legal Warfare activities in peacetime and war.

1-59. <u>Psychological Warfare</u> is broadly similar to U.S. military information support operations in that it is intended to influence the behavior of a given audience. PLA Psychological Warfare seeks to integrate with conventional warfare and includes both offensive and defensive measures. The PLA views Psychological Warfare through the lens of Sun Tzu, emphasizing its multiplicative effect when coupled with comprehensive deception operations. Deception operations are critical to the PLA's entire warfighting approach, and Psychological Warfare represents the major information operations element of deception operations.

1-60. <u>Legal Warfare</u> refers to setting the legal conditions for victory—both domestically and internationally. The U.S. does not have an equivalent concept, although State Department diplomatic and legal operations have roughly equivalent objectives. Legal Warfare seeks to unbalance potential opponents by using international or domestic laws to undermine their military operations, to seek legal validity for PLA operations worldwide, and to support Chinese interests through a valid legal framework. Legal Warfare has emerged with a particularly prominent role via the various Chinese political maneuverings in the Western Pacific, particularly those areas surrounding international waterways, disputed land masses, and economic rights of way. Legal Warfare is present at the tactical and operational levels of war. It guides how the PLA trains to treat prisoners of war, detainees, and civilians, and how it abides by international legal conventions, codes, and laws.

## FRAMEWORK FOR PLA OPERATIONS

1-61. The PLA operational framework is the lens through which the PLA develops capabilities, plans and executes operations, develops doctrine, and revises military philosophy and strategy. The operational framework is complex and comprehensive, running the entire scope of military operations from national strategy down to small-unit tactics. There is no direct U.S. equivalent, though most of the general ideas are replicated in various ways in U.S. doctrine. The PLA operational framework consists of five levels, ranging from the purely philosophical to the prescriptive and practical. These levels are—

- Military Thought.
- Defense Theory and Defense Doctrine.
- Strategic Principles and Operational Principles.
- Campaigns.
- Combat Tactics and Regulations.

Figure 1-4 illustrates the relationship between the different operational framework levels.

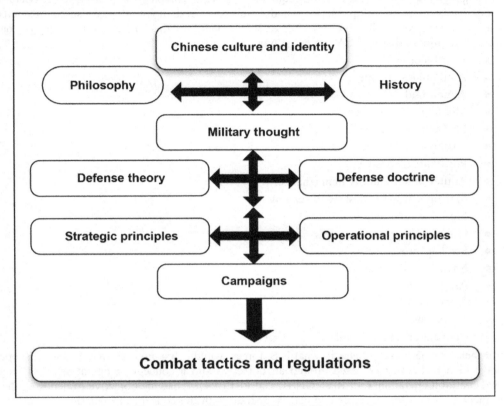

**Figure 1-4. PLA operational framework**

1-62. <u>Military Thought</u> represents the highest levels of PLA military thinking. It embodies China's long history as a land of philosophers, and it contains those works held in the highest esteem by the PLA. <u>The Art of War</u> and People's War are the two foundational documents and concepts for Military Thought, though there are many other works constantly being produced, debated, and revised. Military Thought provides the philosophical foundation for PLA operations, defines the culture and objectives of the PLA, and defines the PLA's relationship with the CPC. U.S. equivalents are military concepts and strategic directives, such as the *National Security Strategy* and the *Capstone Concept for Joint Operations*.

1-63. <u>Defense Theory</u> and <u>Defense Doctrine</u> represent the PLA's strategic-level thinking. Defense Theory operationalizes the various principles laid out in Military Thought and other high-order documents. Defense Doctrine codifies the basic PLA approach to warfighting. Doctrine in this sense remains at the strategic level—the PLA does not appear to use the word "doctrine" to describe operational practices at operational or tactical levels of war. A U.S. equivalent to Defense Theory is the *Air Force Future Operating Concept*; an equivalent to Defense Doctrine is the *U.S. Army in Multi-Domain Operations* (formerly the Army Operating Concept).

1-64. <u>Strategic Principles</u> and <u>Operational Principles</u> align roughly with what the U.S. DOD refers to as doctrine; in other words, a series of situation-based discussions that serve as a guide to actions. These principles are intended to standardize language and understanding throughout the PLA. One important difference between Western doctrine and Chinese Strategic and Operational Principles is the importance that the PLA places on the interactions and relationship between the military and society. Western militaries are generally apolitical and consider themselves separate entities from civilian society, whereas the PLA codifies its place in Chinese society in these principles. In practice, this establishes the role of the PLA in shaping Chinese society and politics, explains the relationship between militias and the areas they serve, and describes the role of political officers in PLA units. A U.S. equivalent to Strategic Principles is JP 3-0, while an equivalent to Operational Principles is ADP 3-0.

1-65. Campaigns do not have a Western equivalent. A PLA campaign is a large-scale, operational-level effort that likely involves multiple services and a joint command, though it may also be conducted by units from a single service. Unlike Western theater-level operations that are generally objective-based, PLA campaigns are mission-based. PLA campaign types include—

- Joint blockade.
- Amphibious landing.
- Antiair raid.
- Mobile/maneuver warfare.
- Mountain offensive.
- Positional offensive.
- Positional defensive.
- Antiterrorism stability maintenance.
- Maritime group to destroy the enemy.
- Maritime interdiction.
- Offensive against coral reefs.
- Sea lane protection.
- Naval base defense.
- Air offensive.
- Airborne.
- Air defense.
- Conventional missile attack.

Each of these campaigns correlates to what U.S. forces would call a *mission*. In contrast, U.S. operations at this level tend to be largely based around an objective and a concept of operations. For example, the U.S.-led coalition operation, OPERATION IRAQI FREEDOM, had the objective of deposing the Ba'athist government and installing a stable democracy in Iraq, with the concept of operations being to defeat the Iraqi military, secure civil stability, and transition to a new government. The PLA's campaign approach suits its doctrine and philosophy well—it is more prescriptive, centralized, and narrow in scope, and consequently it requires less freedom of action from subordinates. It remains to be seen if this approach will be retained as the PLA moves toward more decentralized leadership and actions.

1-66. Combat Tactics and Regulations address unit-level activities and tactical-level doctrine. Tactics and Regulations provide the basis for campaign operations, ensuring that all PLA units have a shared understanding of basic maneuvers, actions, and responses. Regulations in this context are not laws and rules as they are for the U.S. military; instead, they are similar to battle drills or other basic military actions. This mirrors the old naming convention for U.S. Army doctrine, as predecessors to ADP 3-0 and FM 100-5 were titled *Field Service Regulations, Operations* until the mid-1960s. This naming convention may also reflect greater centralization and top-down leadership. This approach may change in a similar way to the campaign model as the PLA moves toward a decentralized approach to leadership.

## PLA SYSTEM WARFARE

1-67. The PLA appears to recognize domains in the same way as the U.S. DOD, and it is actively seeking to enhance multi-domain capabilities and cross-domain integration. The organization applies all of its basic military philosophies and principles of active defense, deterrence, and deception to operations in all domains. The overarching PLA theoretical framework for its multi-domain effort is called system warfare. The system warfare concept seeks to identify critical or vulnerable system components, then degrade or destroy the effective use of larger systems through targeted attacks on these vulnerabilities. The primary principle of system warfare is the identification and isolation of critical or vulnerable subsystems or interdependence of these subsystems. The PLA believes that if key threat systems are rendered ineffective, the threat's ability and will to resist will crumble.

1-68. System warfare is the most recent PLA effort to operationalize the principles of force concentration and asymmetric attack as outlined by Sun Tzu and Mao, while accounting for the lethality, cost, and complexity of modern weapons systems. The PLA classifies all capabilities, ranging from ballistic missiles and strike fighters to cyber operators and special operations forces, as military systems. Each system has inherent strengths and weaknesses. System warfare involves—

- Bypassing enemy systems' areas of strength, thus gaining a combat advantage by approaching them asymmetrically.
- Developing systems that excel at exploiting perceived weaknesses in enemy systems, thereby offsetting their strengths by undermining the systems' ability to perform assigned missions.

The most common examples of system warfare are actions such as targeting networks instead of shooters, sensors instead of aircraft, or command and communication nodes instead of maneuver forces. The PLA expands system warfare to include diplomatic efforts undermining international alliances, offensive cyber operations disabling air or seaport operations, or special operations forces undermining civilian morale through covert operations.

1-69. Many of the systems with which the PLA intends to prosecute system warfare are not standing capabilities, but rather purpose built in times of conflict. During times of war, the PLA intends to build task-organized suites of capabilities designed to strike specific weak points of its opponent's key systems. These suites of capabilities are called operational systems. Each operational system consists of five main subcomponents: the command system, the strike system, the information warfare system, the intelligence system, and the support system.

1-70. At the tactical level, system warfare centers largely on targeting high-value battlefield systems such as radars, command and communication nodes, and field artillery and air defense systems, and it can include selective armored vehicles and critical logistics support means. Examples of tactical system warfare include using heavy rocket artillery to defeat or destroy enemy radars and artillery systems, EW to suppress or neutralize enemy command and communication networks, and deception operations to target enemy leadership's situational understanding and state of mind. Tactical system warfare is discussed in greater detail in chapter 4.

1-71. The PLA's employment of system warfare supports the development of several traditional military strategies, such as preclusion, isolation, and sanctuary, throughout all domains and at all levels of war. Preclusion is achieved by keeping enemy commanders and forces off balance through asymmetric means, such as deception and information warfare, while simultaneously denying use of wide geographic areas through long-range reconnaissance-strike capabilities. Isolation is achieved by jamming or manipulating communications between units, employing psychological warfare to confuse and segregate enemy units from one another, then rapidly maneuvering to physically isolate them. Sanctuary is achieved through a mix of protection, defensive planning, information warfare, and deception operations. Sanctuary includes not only safety from physical attack, but safety from enemy information operations.

This page intentionally left blank.

# Chapter 2

# People's Liberation Army Force Structure

This chapter outlines the organization of the People's Liberation Army (PLA), beginning with its organization at the highest levels of the Communist Party of China (CPC), then describing unit-level organizations at the strategic, operational, and tactical levels. The PLA is one of the world's largest formal organizations, and by military standards it has a very complex organizational structure. Ongoing reforms and reorganization add to this complexity. Diagrams and descriptions in this section provide generalized data, and unit organizational diagrams are examples of probable task-organized capabilities adapted to a mission set.

## CHINA'S SECURITY APPARATUS

2-1.    The PLA represents the armed forces branch of China's security organizations. Other organizations under this umbrella include the Ministry of Public Security (MPS), the Ministry of State Security (MSS), the People's Armed Police (PAP), and the China Militia. These organizations are collectively referred to as the Chinese Security Apparatus. The Chinese Security Apparatus is not fully analogous to the U.S. Department of Defense (DOD): it is more of a descriptive term than a discrete organization. All Security Apparatus organizations fall under the purview of the CPC, and their missions often overlap with one another. One of the defining elements of Chinese internal politics is the complex interplay between these organizations and their leadership. In most cases, the PLA can be thought of as the most important and most influential member of the Chinese Security Apparatus.

2-2.    Forces making up each organization in the Chinese Security Apparatus can be described as either civilian, paramilitary, or military. The composition of each organization varies significantly. Police, security, and some intelligence forces are described as civilian; the MPS and MSS are comprised mostly of these personnel. The PAP and the China Militia are largely paramilitary: they conduct military-like training and employ some military equipment and tactics, but they are not considered part of the PLA.

2-3.    The MPS is China's national police force and the world's largest police organization. Though a national-level force, it is organized and commanded locally, typically by provincial governments. It is responsible for day-to-day law enforcement throughout China, and it is directly reportable to China's State Council. The MSS is China's primary national intelligence apparatus, having responsibility for both foreign and domestic intelligence work and counterintelligence. The MSS's counterintelligence mission requires a fairly significant "secret police" element that does not work with the MPS. The PLA generally does not integrate activities with either the MPS or the MSS, except when intelligence or security information is exchanged. The roles and responsibility of the MSS are in transition, as a new high-level command—the People's Liberation Army Strategic Support Force (PLASSF)—was recently established and appears to overlap numerous responsibilities with the MSS.

2-4.    The Chinese Armed Forces consist of the PLA, the PAP, and the China Militia. The PLA's primary role within the Chinese Armed Forces is to defend China from external or international threats, while the PAP focuses on internal threats. The PAP's mission is internal security and, in conjunction with the MPS, limited law enforcement. The PAP also has the official mission to support or reinforce the PLA during times of war. PAP forces are organized on a provincial basis, with two national-level antiterrorism units stationed in the capital region. The PAP features military-like organizations and extensive use of military equipment, including small arms, helicopters, and light armored vehicles. The U.S. does not have an equivalent organization.

2-5.   The China Militia is a massive quasi-formal militia element of the Chinese Armed Forces, consisting mostly of poorly trained and equipped part-time regional military units. The militia's primary mission is to provide logistics and security support to the PLA, though disaster relief and internal security missions are also part of its mission set. The China Militia is a different entity from the People's Liberation Army Reserve. The militia can be can be considered an aspect of People's War philosophy, to be integrated with the military and other entities in times of national emergency or crisis. Like the rest of the Chinese Armed Forces, it is being downsized and modernized, with less emphasis put on manpower and more emphasis on quality training and equipment. Like the PAP, there is not a true U.S. equivalent to the China Militia.

2-6.   The PLA is the armed forces of China. Though called an army, the PLA is analogous to the U.S. DOD in that it consists of all of the service branches: an army (People's Liberation Army Army—PLAA), navy (People's Liberation Army Navy—PLAN), air force (People's Liberation Army Air Force—PLAAF), rocket force (People's Liberation Army Rocket Force—PLARF), and strategic support force (PLASSF). In addition, the People's Liberation Army Joint Logistics Support Force (PLAJLSF), though not a service, is a national-level PLA organization. The PLA is under the command of the Central Military Commission (CMC), as supervised by the Politburo Standing Committee and the CPC. The PLA is officially the armed wing of the CPC; all levels of the PLA are supervised by a system of political officers. The expressed primary missions of the PLA are to protect the ruling status of the CPC, ensure China's sovereignty and territorial integrity, safeguard Chinese interests at home and abroad, and help maintain global stability. (See figure 2-1 on page 2-3 for a simplified diagram of the PLA's command structure.)

2-7.   The PLAA is China's land combat service. It consists of an active-duty component and a reserve component, organized in roughly the same way as the U.S. Army's active and reserve components. The PLAA's active force consists of roughly one million soldiers, and it remains the dominant service within the PLA, despite recent efforts to reduce its influence. The reserve component consists of roughly 500,000 soldiers and, despite being a national-level organization, it is relatively poorly trained and equipped—though it is a significant improvement over the China Militia. In times of war, the reserve component will likely take over rear area, security, and sustainment duties in order to free up higher-quality active duty forces for more rigorous missions. The PLAA recognizes branches in much the same way the U.S. Army does. PLAA branches are—

- Infantry (including mechanized, motorized, and mountain).
- Armored.
- Artillery (including towed and self-propelled tube artillery, rocket artillery, antitank missiles and guns).
- Air defense (including short- and medium-range mobile gun and missile systems—heavier missile systems are operated by the PLAAF).
- Aviation (including helicopters and limited fixed-wing aircraft).
- Engineer.
- Chemical defense.
- Communications.
- Electronic warfare (EW).
- Logistics.
- Armaments (maintenance).
- Special operations forces (SOF).

# CHINESE NATIONAL AND STRATEGIC-LEVEL ORGANIZATIONS

2-8.   The command structure of the PLA is complex and opaque to outsiders. Both its command structure and apparatus are deliberately complex, designed first and foremost to ensure the security of the CPC. In stark contrast to Western military philosophy, Chinese command structures do not generally appoint a single decision maker or imbue a single individual with absolute command authority: decisions are to be made via consensus, rather than by fiat. Of note, however, the current Chinese President, Xi Jinping, appears to wield much greater direct authority over the PLA than his predecessors. There are no fewer than ten different national-level command organizations in China, organized across at least three different levels of a complex hierarchy. This ATP will not discuss in detail the intricate web of organizations and internal politics at the top of China's command structure, but it will provide a simplified overview to enable better understanding of how these command relationships influence tactical operations.

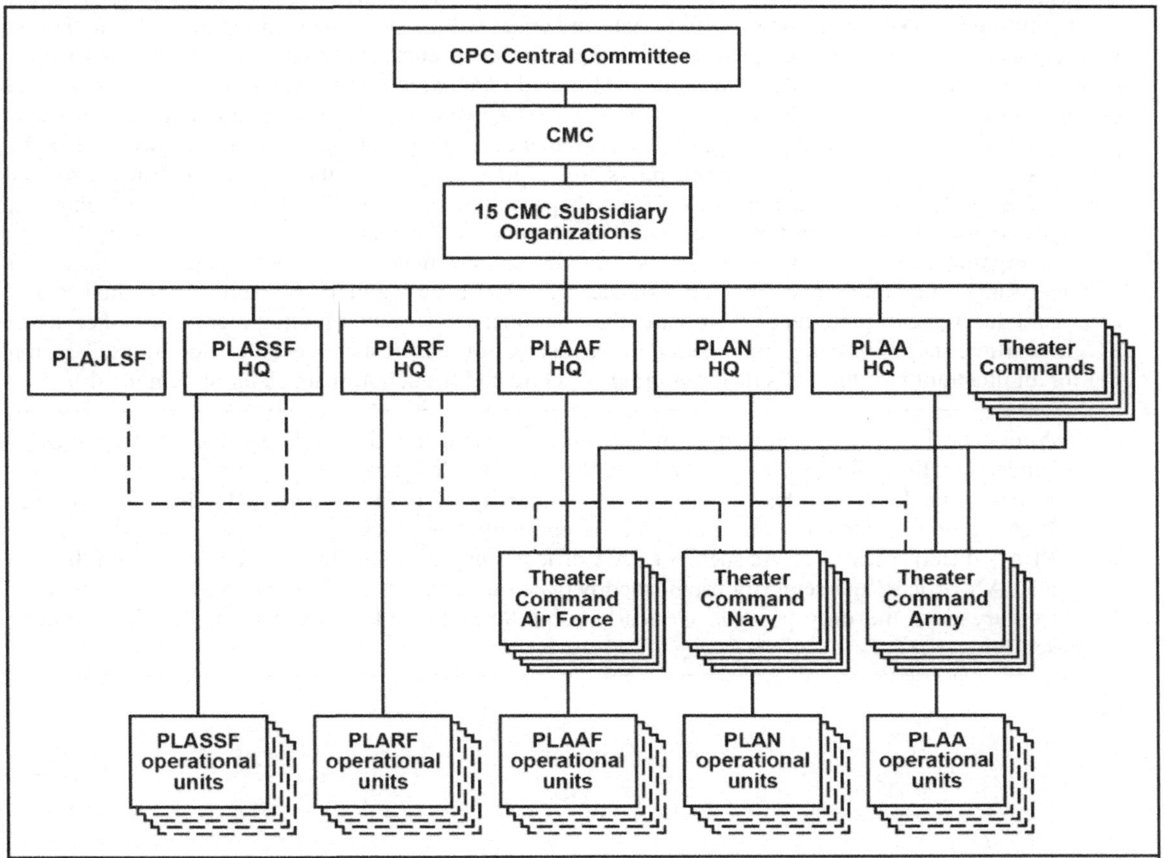

**Note.** Theater command branches report to two entities: their respective theater commands and their headquarters. For example, the Western Theater Command Army reports to the Western Theater Command and PLAA HQ.

| | | | |
|---|---|---|---|
| CMC | Central Military Commission | PLAJLSF | People's Liberation Army Joint Logistics Support |
| CPC | Chinese Communist Party | PLAN | People's Liberation Army Navy |
| HQ | headquarters | PLARF | People's Liberation Army Rocket Force |
| PLAA | People's Liberation Army Army | PLASSF | People's Liberation Army Strategic Support Force |
| PLAAF | People's Liberation Army Air Force | ——— | command relationship |
| | | – – – – – | support relationship |

**Figure 2-1. Simplified PLA command structure**

2-9. The CMC is the most powerful and significant national-level command body in China. It is also the most relevant to tactical operations. Recent restructuring created no fewer than 15 functional departments, including a national-level army department and a strategic rocket department. In general, national headquarters are responsible for capabilities development, education, and training, while theater commands (discussed in paragraph 2-10) are responsible for operations. The CMC is notionally led by a civilian chairman: the CPC's head and China's president. He is typically assisted in decision making by numerous vice chairmen and senior PLA officers. The CMC's command structure relies on a committee approach and is not formal—it routinely changes based on political dynamics. This committee-based approach differs significantly from the U.S. approach, where a single decision maker—the President—has all command authority and is only advised by his staff and cabinet. The Chinese approach likely arose from the desire to make all military decisions with CPC political objectives in mind, and it helps explain why political officers are present throughout the PLA chain of command.

2-10. CMC command authority is exercised operationally through <u>theater commands</u> (TCs). TCs are joint commands that exercise extensive—though not exclusive—command authority in their given region. TCs are tasked with developing strategy, tactics, and policy specific for their areas of responsibility, and they are directly responsible for responding to threats and crises within their assigned regions. TCs are similar to U.S. combatant commands, except that the parallel chains of command extend to the national command structure. TCs are strictly domestic in scope, and they do not extend past Chinese land borders and nearby maritime regions. TC commands mimic Chinese national command structure in that they are heavily politicized and bureaucratically complex. TC commanders share command responsibility with political commissars of the same rank; TC commanders are responsible for operations, while political commissars are responsible for ideological functions. TC staffs participate in committee-based decision making. Each TC has a single-service subordinate headquarters responsible for that service's operations in the theater, plus an organic electronic countermeasures brigade, an information operations support brigade, and a reconnaissance and intelligence support brigade. PLAA, PLAAF and PLAN units have dual chains of command during peacetime, reporting both to their national service-level headquarters and to their assigned TC. In wartime they will likely be under complete control of their respective TCs. (See figure 2-2 on page 2-5 for an illustration of the TC's dual reporting structure and the subordinate chain of command.)

---

***Note.*** The PLA understands rank in the same way that the U.S. military does, but it takes a fundamentally different view of grade. In the PLA, grade refers to one's duty position: platoon leader, brigade deputy leader, and so on. Hierarchy of grade is extremely important to the PLA—more so than rank. There are 15 different grades, and they are recognized across all of the PLA's different services. As such, a PLAA officer's grade can be directly compared with that of a PLAN officer, and seniority established between the two. In the U.S. military, grade refers to military pay, and duty position is clearly subordinate to rank when it comes to determining seniority.

---

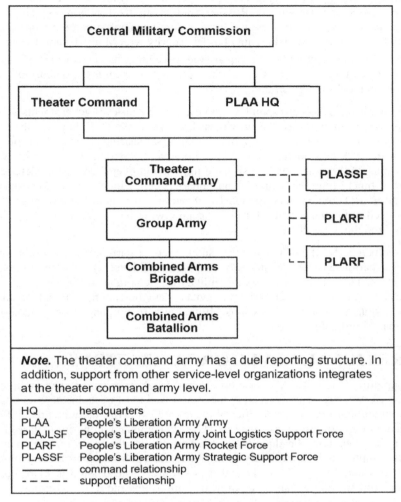

**Figure 2-2. PLAA chain of command, isolated**

2-11. TCs were reorganized in 2016, reducing the number of regions from seven to five and changing the title of <u>military region</u> to TC. When in a list, TCs are presented in order of protocol, with the most prestigious—and most militarily relevant—TC listed first. The current TCs, in order of precedence, are—

- <u>Eastern Theater Command</u>, headquartered in Nanjing, with responsibility for central eastern China and the East China Sea, including the Strait of Taiwan. The Eastern TC likely has operational responsibility for matters involving Taiwan, Japan, and anything related to the East China Sea.
- <u>Southern Theater Command</u>, headquartered in Guangzhou, has responsibility for south-central China, including the border with Vietnam and the South China Sea. The Southern TC's primary missions are maintaining security in the South China Sea and supporting the Eastern TC in any major amphibious operation.
- <u>Western Theater Command</u>, headquartered in Chengdu, has responsibility for virtually the entire western half of China, including borders with India and Russia and the regions of Tibet and Xinjiang. The Western TC's primary missions are contending with perceived separatist and terrorist threats in Tibet and Xinjiang and addressing border issues with China's two powerful western neighbors.

- <u>Northern Theater Command</u>, headquartered in Shenyang, with responsibility for northeastern China along the Mongolian, Russian, and North Korean borders. The Northern TC's most prominent responsibilities are the various contingencies stemming from China's shared border with North Korea and maintaining border security with Mongolia and Russia.
- <u>Central Theater Command</u>, headquartered in Beijing, has responsibility for north-central China and the capital region. The Central TC's primary mission is the defense of Beijing, and it serves as the national strategic military reserve.

2-12. Military districts (MDs) correspond to Chinese provincial-level governments and are considered the military organs within their respective provincial-level governments. They use the same dual-command structure as TCs, with a commander and political commissar sharing command responsibility. The primary responsibilities of MD headquarters are conscription and demobilization, Military-Civil Fusion, and command of China Militia units. MDs also have jurisdiction over PLA border defense units in Xinjiang, Tibet, and part of Inner Mongolia, while TC headquarters command border and coastal defense units in the rest of the country. MD commanders previously carried the grade equivalent of a corps leader, but this has likely been reduced following recent reforms in keeping with the general theme of professionalization and modernization throughout the PLA.

2-13. Local commands (LCs), such as military subdistricts or garrisons at the prefecture, county, and city level, and PLA departments at the county, city, and municipal level represent the lowest level of the PLA's command structure. They are described as the military service organs of the people's government. LCs are primarily charged with meeting conscription quotas, demobilization, national defense education, local Military-Civil Fusion, and command of China Militia units. In Xinjiang, Tibet, and part of Inner Mongolia they also command border defense units.

# PLAA OPERATIONAL-LEVEL ORGANIZATIONS

2-14. The <u>group army</u> is the PLAA's basic operational-level organization. The group army structure is an evolution of the PLA's corps-based structure that comprised most of its history; it appears the group army is an attempt to retain most of the capabilities of the traditional corps, but with greater flexibility and ability to task-organize. Group armies are assigned to TCs; TC command authority is passed through TC PLAA headquarters to the group army headquarters. Group armies use the same dual-command structure as most other PLA units, employing both a military commander and a political commissar. Compared to geographic headquarters (MDs and LCs), it is likely that a group army's leadership is more heavily influenced by operational requirements, rather than local political relationships and responsibilities.

2-15. Following a complete overhaul and reorganization in 2017, each group army now directly commands 12 brigade-size organizations: six combined-arms brigades and six support brigades of various types. Except for six legacy divisions, divisional headquarters have been eliminated, notionalized, or made strictly administrative. This overhaul seeks to eliminate excess command structure, reduce the number of general officers and associated staffs, and increase capabilities at tactical-level formations. The group army reorganization coincided with a massive drawdown in manpower across the PLA. This transition seeks to increase combined arms capability across the PLA through the application of improved technology and training, while simultaneously reducing PLA manpower and reducing quantities of obsolete equipment. A more standardized group army is a centerpiece of this effort.

2-16. The group army is likely not intended to be employed as an operational unit. Rather, it is the force pool from which operational systems are built as part of the wider system-warfare construct. Group army commanders facilitate the assembly of purpose-built operational systems, using their available force structure to create the command, maneuver, and support systems that execute operations in the group army's combat area. As such, concerns about the group army's large number of subordinate units—and the ability, or lack thereof, to control them—are not reflective of the PLAA's approach to building forces.

2-17. The main combat power of the notional group army consists of its six combined arms brigades (CA-BDEs). These brigades are supported by one artillery brigade; one air defense brigade; an aviation brigade; an SOF brigade; an engineer and chemical defense brigade; and a service support brigade, consisting of logistics, transportation, medical, repair, ammunition, communications, unmanned aircraft systems (UAS), and EW units. The 2017 reorganization placed a greater emphasis on system warfare capabilities at the group army level, providing a much more extensive suite of EW and cyber capabilities, long-range reconnaissance, and long-range fires under the direct control of group army commanders. Group army commanders can now support their assigned CA-BDEs with a significant suite of capabilities able to influence operations across all domains. (See figure 2-3 for a graphic depiction of the group army.)

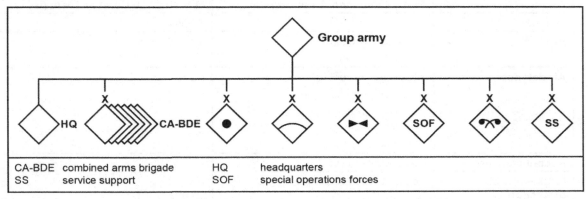

**Figure 2-3. Group army structure (doctrinal)**

2-18. Group-army joint capabilities are limited, but they are expected to expand with ongoing reforms. TCs provide air, naval, and some SOF support, though group armies do typically contain an SOF brigade. The group army also appears to be largely dependent on higher commands—either the TC or PLAJLSF—for most of its logistics support. This emphasizes the lack of expeditionary capability throughout the PLA: sustainment within China's borders may be sufficient to support high-intensity operations, but current sustainment capabilities likely cannot support major combat operations overseas.

# PLAA TACTICAL-LEVEL ORGANIZATIONS

2-19. The PLAA has completed a substantial period of reorganization with regard to its tactical unit structure. Traditionally, the PLAA was built around the division. Divisions were designed around the Soviet model. They were somewhat smaller and more homogenous than their U.S. counterparts, without the same combined arms and sustainment capabilities. In a transition similar to that undertaken by the U.S. Army, Ground Forces of the Russian Federation, and numerous other armies, the PLAA is in the process of "brigade-ization"—moving capabilities that used to reside at the division echelon to the brigade, with the goal of creating a more flexible force with a streamlined command structure, from group to army to brigade to battalion. This reorganization corresponds with the substantial drawdown in manpower ongoing throughout the PLA. PLAA divisions are now largely an administrative echelon—only six divisions remain operational, with none residing in group armies. Otherwise, frontline PLA units have moved entirely to a brigade-oriented structure.

> *Note.* Unit descriptions in this section are notional, as described by the PLAA and supported by further intelligence analysis. Real-world units may vary widely in composition, organization, and equipment, though the PLAA is moving in the direction of force-wide standardization. It is important to note that PLAA units at battalion and below are designed to fight as structured, without the need for significant task organization.

## COMBINED ARMS BRIGADES

2-20. The CA-BDE is the PLAA's basic operational unit. It is similar in size, capability, and organization to the U.S. Army's brigade combat team (BCT); it is entirely possible that the PLAA was influenced by BCT organization when designing the CA-BDE. The CA-BDE mixes different capabilities under a single headquarters: maneuver, fires, logistics, communications, engineer, and EW subordinate units are organically assigned to CA-BDE headquarters. CA-BDEs do not appear to have any organic joint capability, but they are developing joint capabilities to be able to control units from other services during training and operations. (See table 2-1 for a comparison of a PLAA CA-BDE to a U.S. BCT.)

**Table 2-1. Comparison of PLAA combined arms brigade to U.S. brigade combat team**

|  | *PLAA Combined Arms Brigade* | *U.S. Brigade Combat Team (BCT)* |
|---|---|---|
| **Maneuver** | 4–6 battalions | 3 battalions |
| **Artillery** | 1 howitzer battalion<br>1 rocket battalion | 1 howitzer battalion |
| **Air defense** | 3 batteries, including:<br>• Self-propelled guns.<br>• Man-portable air defense systems.<br>• Short-range missile systems. | None |
| **Reconnaissance** | Comparable to U.S. BCT | |
| **Engineer and protection** | Comparable to U.S. BCT | |
| **Logistics and sustainment** | Newly established<br>Likely less capable than U.S. BCT | |

2-21. There are three distinct types of CA-BDEs: light (motorized), medium (mechanized), and heavy (armored). The PLAA describes the differences between motorized and mechanized infantry in how supporting vehicles are employed: motorized units are only transported by their assigned vehicles, while mechanized forces employ their vehicles as combat platforms that support the infantry. The PLAA employs a variety of armored personnel carriers (APCs) and infantry fighting vehicles (IFVs) that feature a broad range of firepower and protection; some are tracked, some are wheeled, and there is considerable overlap. As such, one must look at how the unit intends to fight, rather than its composition and equipment, when assessing a unit as motorized versus mechanized. Airborne, mountain, and amphibious CA-BDEs are described as light. The CA-BDE organizations are described in paragraphs 2-22 through 2-24.

2-22. The light CA-BDE contains these units—
- Four motorized combined arms battalions (CA-BNs).
- One reconnaissance battalion.
- One artillery battalion.
- One air defense battalion.
- One headquarters unit.
- One operational support battalion.
- One service support battalion.

(See figure 2-4 for a graphic depiction of the light CA-BDE.).

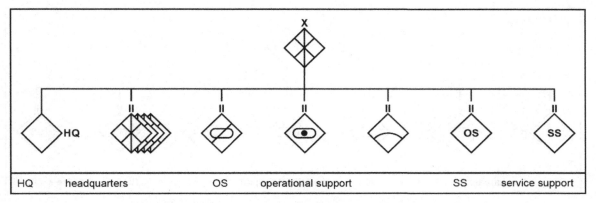

**Figure 2-4. Light combined arms brigade (doctrinal)**

2-23. The medium CA-BDE contains these units—
- Four mechanized CA-BNs.
- One reconnaissance battalion.
- One artillery battalion.
- One air defense battalion.
- One headquarters unit.
- One operational support battalion.
- One service support battalion.

(See figure 2-5 for a graphic depiction of the medium CA-BDE.)

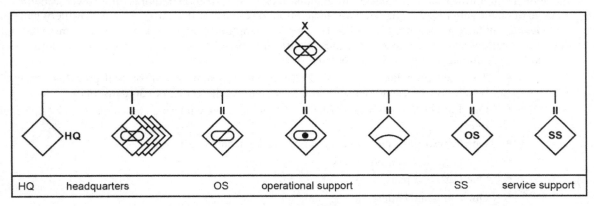

**Figure 2-5. Medium combined arms brigade (doctrinal)**

2-24. The heavy CA-BDE contains these units—
- Four armored CA-BNs.
- One reconnaissance battalion.
- One artillery battalion.
- One air defense battalion.
- One headquarters unit.
- One operational support battalion.
- One service support battalion.

(See figure 2-6 for a graphic depiction of the heavy CA-BDE.)

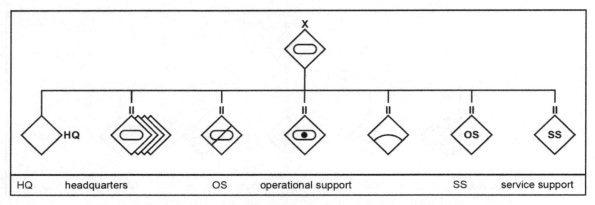

**Figure 2-6. Heavy combined arms brigade (doctrinal)**

## ARTILLERY BRIGADE

2-25. The artillery brigade in the group army employs a variety of towed guns, self-propelled guns (SPGs), light (122-mm) and heavy (300-mm) rocket artillery systems, and antitank and assault vehicles. These systems are employed to mass fires on critical targets, reinforce fires at lower echelons (chiefly the CA-BDE), deter or deny enemy actions, and offset enemy advantages in close combat. Artillery brigade assets may be employed in direct or general support of CA-BDEs. Artillery battalions include organic surveillance and target acquisition assets, including UAS, electronic intelligence (ELINT) systems, and traditional long-range visual forward-observation platforms. It is not clear how effectively the PLAA can task-organize fires; it traditionally preferred to centralize fires in order to maximize the effects of mass, but the movement toward smaller tactical formations and modularization requires that lower echelons be capable of employing effective fire support. In addition to its indirect fire capability, each artillery battalion includes an antitank guided missile (ATGM) company, employing light armored vehicles mounted with ATGMs. Artillery brigade composition varies significantly based on operational requirements and system availability. A notional artillery brigade contains—

- Two self-propelled 122-mm, 152-mm, or 155-mm towed or self-propelled howitzer battalions (three batteries with four to six guns each, from 24 to 36 guns total).
- One light (122-mm) rocket battalion (three batteries with nine launchers each, 27 launchers total).
- One heavy (300-mm) rocket battalion (12 launchers total).
- One target acquisition battery.
- One UAS company.
- One command battery.
- One support company.

(See figure 2-7 for a graphic depiction of the artillery brigade.)

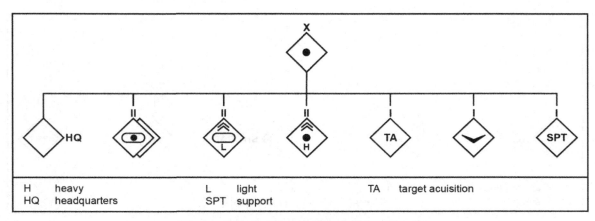

**Figure 2-7. Artillery brigade (doctrinal)**

## AIR DEFENSE BRIGADE

2-26. The group army's air defense brigade provides the middle tier in the PLA's tiered and layered approach to ground-based air defense. The PLAA historically relied heavily on antiaircraft guns, both towed and self-propelled (SPAAGs). These systems are still heavily represented in both the air defense brigade and in the air defense battalions of the CA-BDEs. The air defense brigade also employs a mixture of medium-range radar missile systems, short-range missile systems, and sensors designed to support engagement of aerial targets. It is not known to what extent these tactical air defense systems integrate with the much-larger theater-wide air defense networks employed by TCs, but it is likely that at least the medium-range surface-to-air missile (SAM) battalion has some sort of digital integration with a higher air defense echelon. As with artillery systems, air defense units may be employed in mass to defend critical assets from air attack, or they may be detached in direct or general support to subordinate CA-BDEs. Each brigade may also include an electronic air defense battalion, integrating EW assets with gun systems. As with artillery systems, air defense brigade composition varies significantly based on operational requirements and system availability. A notional air defense brigade contains—

- One medium-range SAM battalion (three batteries consisting of one radar section and three launcher sections per battery).
- Two short-range air defense (SHORAD) battalions (three batteries consisting of six to eight SPAAGs and four to six short-range SAM launchers per battery).
- One electronic air defense battalion (three batteries consisting of varied EW capabilities).
- One command battery.
- One support company.

(See figure 2-8 on page 2-12 for a graphic depiction of the air defense brigade.)

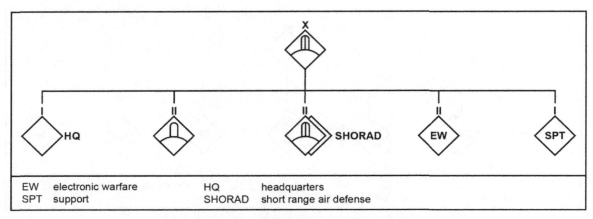

**Figure 2-8. Air defense brigade (doctrinal)**

## ENGINEER AND CHEMICAL DEFENSE BRIGADE

2-27. The group army's engineer and chemical defense brigade is responsible for protection, mobility, countermobility, and chemical defense and obscuration functions within the group army's combat area. Historically, the PLA invested heavily in its engineer units due not only to its emphasis on mobility and countermobility, but also for contingency missions in support of disaster relief or other civil emergencies. This employment strategy is likely to continue. PLAA engineer brigades include mining, countermining, obstacle, bridging, smoke, chemical defense, and heavy maintenance units. These units provide much of the PLA's support to international peacekeeping and development missions.

## SERVICE SUPPORT BRIGADE

2-28. The service support brigade includes logistics, transportation, medical, repair, ammunition, command and communication, UAS, and EW units. In addition to traditional combat support roles, the brigade is postured to perform signal support and EW missions. The service support brigade's signal element employs both traditional and network communications throughout the group army's combat area. Previously, communications were handled largely at the division, with minimal communications architecture being employed at lower echelons. With the PLAA's emphasis on information operations, signal capabilities were significantly enhanced at tactical echelons. The signal element has responsibility for providing network connectivity to PLAA units and protecting these communications networks from jamming, electromagnetic attack, and cyber intrusion. The EW regiment represents a serious investment in EW and cyber capability to be employed at the operational and tactical levels. It includes a jamming and electromagnetic attack section, a long-range electronic surveillance section, an electromagnetic protection section, a network operations section, and a communications operations section. It is unclear how the PLA will allocate permissions and responsibilities to the EW regiment, though it can be safely assumed that rules of engagement will be more permissive than they are in Western militaries. It can also be assumed that the EW regiment will work closely with the artillery and air defense brigades in order to target long-range artillery fires in a construct similar to the Russian reconnaissance-fire system.

*Note.* The PLAA describes signal and communications capabilities under the umbrella of command and communication.

## ARMY AVIATION BRIGADE

2-29. The army aviation brigade (AAB) provides the group army's rotary-wing aviation capability. The AAB's development is likely influenced by the U.S. Army's combat aviation brigade, though the AAB has far fewer airframes than does its U.S. counterpart. The AAB is built around medium-lift transport helicopters and scout and attack helicopters. The PLAA uses a number of different helicopter types: foreign systems from Russia, the former Soviet Union, and Europe, and indigenous designs are all present in significant numbers. The density of helicopters in the PLAA is relatively low as compared to the size of the rest of the force; the U.S. Army's ratio of helicopters to troops is over ten times that of the PLAA. Rotary-wing capability is a significant area of investment in the PLA's modernization strategy; it is likely that the AABs will be greatly expanded in the relatively near future, using mainly indigenous or license-produced airframes. The AAB is organized into eight battalions, mixing general-utility lift helicopters, attack helicopters, and light attack and reconnaissance helicopters. In addition to the AABs, the PLAA has also formed two air assault brigades, which include either two or three infantry battalions and up to six medium-lift helicopter battalions. AABs train frequently with both light CA-BDEs and SOF. A notional AAB organization contains—

- Four general-utility battalions (eight to 12 medium-lift or general-utility helicopters per battalion).
- Two attack battalions (eight attack helicopters per battalion).
- One reconnaissance battalion (eight reconnaissance or light attack helicopters).
- One headquarters unit.
- One aviation support battalion.

(See figure 2-9 for a graphic depiction of the army aviation brigade.)

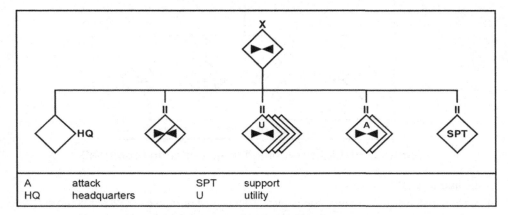

**Figure 2-9. Army aviation brigade (doctrinal)**

## SPECIAL OPERATIONS FORCES BRIGADE

2-30. The SOF brigade provides the group army with an organic SOF capability. Group-army SOF brigade operations generally do not focus on training or interacting with foreign militaries; they instead focus on direct action—deep reconnaissance and commando operations in support of group army operations. Group army SOF brigades are highly specialized to operate in their specific theater. Mountainous theaters focus on alpine training, coastal theaters focus on amphibious operations, and urban theaters focus on urban operations. SOF brigades are equipped with organic surveillance UASs, likely to aid target acquisition in support of heavy artillery and rocket fires. Group army SOF brigades are more similar to U.S. Army Ranger light infantry and long-range reconnaissance units than traditional SOF; their primary mission is to support the group army's combined arms maneuver operations. SOF brigades habitually train with PLAA aviation units, but they do not enjoy the wide variety of fixed- and rotary-wing capabilities available to U.S. SOF units.

## COMBINED ARMS BATTALION

2-31. The CA-BN is a very new development in the PLAA organization. The CA-BN takes the basic combined arms approach used to build the CA-BDE and applies it to the battalion echelon. CA-BNs appear to only combine different maneuver elements along with organic short-range fires elements (assault guns and mortars), with the provision that CA-BDE headquarters can attach elements from other brigade organizations as required. The CA-BN is very similar to the U.S. battalion task force concept employed by mechanized and armored units since the World War II era, mixing company-level infantry and armor units to create a single combined arms command. Each CA-BN also houses an organic SHORAD capability in the form of man-portable air defense systems (MANPADS). The PLAA breaks CA-BNs into three primary categories: light, medium, and heavy. Of note, the CA-BN appears to have only limited staff, which may affect its ability to function as the PLAA intends—as an independent unit. Combined arms battalion organizations are described in paragraphs 2-32 through 2-34.

2-32. A light infantry CA-BN contains—

- Three motorized infantry companies (10 light wheeled vehicles or APCs per company).
- One firepower company (six to nine rapid-fire 81-mm mortars, the form of man-portable air defense, and crew-served weapons).
- One headquarters unit.
- One service support company.

(See figure 2-10 for a graphic depiction of a light combined arms battalion.)

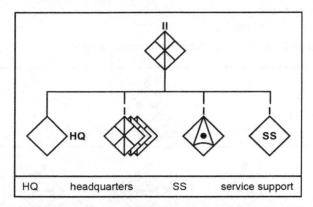

**Figure 2-10. Light combined arms battalion (doctrinal)**

2-33. A medium CA-BN contains—

- Three mechanized infantry companies (10 wheeled or tracked IFVs per company).
- One assault gun company (14 wheeled 105-mm assault guns).
- One firepower company (six to nine rapid-fire 120-mm self-propelled mortars, MANPADS, and crew-served weapons).
- One headquarters unit.
- One service support company.

(See figure 2-11 for a graphic depiction of a medium combined arms battalion.)

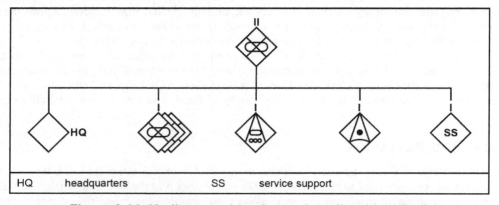

**Figure 2-11. Medium combined arms battalion (doctrinal)**

2-34. A heavy CA-BN contains—

- Two tank companies (10-14 tanks per company).
- Two mechanized infantry company (10 IFVs per company).
- One firepower company (six to nine rapid-fire 120-mm self-propelled mortars, MANPADS, and crew-served weapons).
- One headquarters unit.
- One service support company.

(See figure 2-12 for a graphic depiction of a heavy combined arms battalion.)

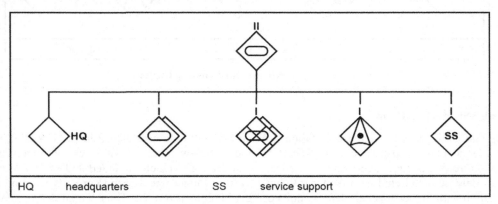

| HQ | headquarters | SS | service support |

**Figure 2-12. Heavy combined arms battalion (doctrinal)**

## ARTILLERY BATTALION

2-35. The CA-BDE's artillery battalion is a major focus of PLAA investment. Recognizing that PLAA forces may be outmatched in close combat with other industrialized nations' maneuver forces, the PLAA developed doctrine designed to defeat, suppress, or neutralize enemy maneuver units before close contact occurs. In practice, this necessitated the fielding of new SPGs in both the 122-mm and 155-mm class, along with a series of older 152-mm towed guns and heavy, highly capable rocket artillery platforms. Unlike the U.S. Army, CA-BDEs employ a composite of rocket and tube artillery at the brigade level. CA-BDEs can be reinforced with guns and rocket artillery from their respective TC's artillery brigade. Of note, the artillery battalion contains a robust antitank capability, consisting of light armored ATGM vehicles and several towed antitank guns. An artillery battalion contains—

- Two to three 122-mm or 155-mm SPG batteries (three platoons with three guns each, 18-27 guns total).
- One light (122-mm) rocket battery (three platoons with three launchers each, nine launchers total).
- One ATGM company (nine vehicle launchers or antitank guns).
- One command battery.
- One support company.

(See figure 2-13 for a graphic depiction of an artillery battalion.)

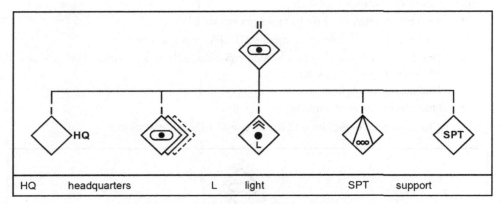

**Figure 2-13. Artillery battalion (doctrinal)**

## AIR DEFENSE BATTALION

2-36. The CA-BDE's air defense battalion provides low-altitude air defense in the CA-BDE's airspace. PLAA tactical air defense systems are generally mobile, lightweight, and within visual range. They consist of a mixture of guns and short-range missiles, including both vehicle mounted and MANPADS systems. The air defense battalion does not appear to have a robust long-range detection or early warning capability; integration with higher-level air defense units or PLAAF units is limited. Thus, the air defense battalion can only provide point, within-visual-range air defense against low-flying targets. The air defense battalion's primary target set is helicopters, unmanned aircraft, and low-flying aircraft. Fires are likely controlled locally through visual aircraft identification and weapons control statuses. Radar detection is limited to short-range organic radars mounted on weapons systems. The air defense battalion's systems are mobile, and they move closely with CA-BDE units while maneuvering. Gun systems are dual purpose and can engage surface targets if necessary. An air defense battalion contains—

- Three SPAAG batteries (six guns per battery, 18 guns total).
- One SHORAD battery (eight systems per battery and possibly one to two radar systems).
- One command battery.
- One support company.

(See figure 2-14 for a graphic depiction of an air defense battalion.)

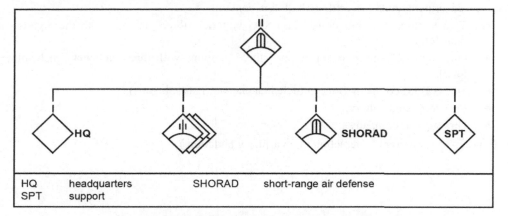

**Figure 2-14. Air defense battalion (doctrinal)**

## RECONNAISSANCE BATTALION

2-37. The CA-BDE's reconnaissance battalion provides multi-domain reconnaissance, intelligence, surveillance, and target acquisition capabilities. The reconnaissance battalion uses a mixture of mounted, dismounted, UAS, human intelligence (HUMINT), ELINT, and cyber collection capabilities to support CA-BDE intelligence and targeting requirements. The reconnaissance battalion's structure is very similar to that of the U.S. Army reconnaissance squadron, consisting of three troops employing a mixture of collection capabilities. In addition to traditional scouts, the reconnaissance battalion may include an element of SOF-like troops, capable of conducting independent direct action in deep areas. The reconnaissance battalion is a critical component of the CA-BDE's reconnaissance-strike capability, providing much of the deep intelligence needed to support targeting for both the CA-BDE's organic surface-to-surface fires and any supporting or reinforcing fires from the artillery brigade. The PLAA places a very high priority on reconnaissance in general. In order to shape the mindset of one's opponents, one must know as much as possible about their disposition and intentions. It should also be noted that group armies do not have a dedicated ground reconnaissance unit as do U.S. corps; the group army's SOF brigade provides much of the ground intelligence in support of group army operations. A reconnaissance battalion contains—

- Two reconnaissance troops (mounted; six to 10 light armored vehicles per troop).
- One battlefield surveillance company (all-source intelligence organization).
- One UAS company (likely containing two to three aircraft).
- One headquarters unit.

(See figure 2-15 on page 2-18 for a graphic depiction of a reconnaissance battalion.)

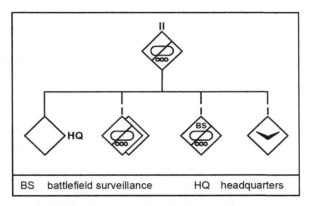

**Figure 2-15. Reconnaissance battalion (doctrinal)**

## OPERATIONAL SUPPORT BATTALION

2-38. The CA-BDE's operational support battalion provides mobility, countermobility, and protection capabilities, along with an EW company, a command and communication company, a chemical defense company, and a security (military police) company. It is similar—though not identical—to the U.S. Army's special troops battalion. The operational support battalion likely operates in a highly decentralized manner, task-organizing support elements to other battalions within the CA-BDE. Operational support battalions are tailored to the needs of their parent unit; heavier CA-BDEs require heavier engineer support. One of the most important missions of the operational support battalion is supporting deception operations. The PLAA puts a very high priority on camouflage and concealment, and much of this responsibility falls to the engineering element of the operational support battalion. The operational support battalion also employs a relatively high density of vehicles, representing a significant portion of the CA-BDE's wheeled-vehicle inventory.

## SERVICE SUPPORT BATTALION

2-39. The CA-BDE's service support battalion provides sustainment support for the CA-BDE, including supply, medical, and maintenance support. It is likely structured similarly to the US Army's brigade support battalion. It is unclear if the service support battalion operates in a more traditional and centralized manner, or if it habitually task-organizes itself to support individual CA-BNs. It can be assumed that, in order to support the CA-BN concept, greater logistics decentralization must occur, but PLAA logistics infrastructure may not support this approach without significant augmentation.

# Chapter 3

# People's Liberation Army Joint Capabilities

This chapter describes People's Liberation Army (PLA) joint capabilities, how they support the tactical commander, and how these forces are integrated into tactical operations. The modern PLA is a sophisticated joint force, likely capable of conducting integrated operations across all domains. PLA joint operations are still in their relative infancy; the PLA has yet to employ a truly integrated joint force in combat, and many key joint integration enablers are still in the developmental stage. Nonetheless, PLA joint operations are likely to improve dramatically as integration technology advances.

## THE PEOPLE'S LIBERATION ARMY AIR FORCE

3-1.   The People's Liberation Army Air Force (PLAAF) is one of the world's largest air forces. Much like the rest of the PLA, the PLAAF is in the midst of a significant reorganization and modernization campaign, moving from employing massive numbers of 1960s-era aircraft with very limited capabilities to employing much smaller numbers of modern 4th-, 4.5th-, and 5th-generation multirole and fighter aircraft. The PLAAF enjoys a suite of capabilities largely comparable to modern Western air forces, including airborne early warning, aerial refueling, heavy and medium transport, and electronic warfare (EW), though relatively few airframes are available for these noncombat missions as compared to the United States Air Force (USAF). The PLAAF also operates a large and advanced ground-based air defense network employing medium- and long-range radar-guided missiles in an integrated, tiered, and layered deployment strategy.

3-2.   For most of the PLAAF's history, its role has been defending Chinese airspace, with secondary importance given to strike and attack missions and little emphasis on close air support (CAS). This made sense for the period, considering the perceived threats to Chinese territory and the lack of available air-ground communications. As the PLAAF has modernized, greater emphasis has been placed on ground support, but these capabilities are developing and are likely limited in both availability and effectiveness. Although many PLAAF airframes are nominally multirole—meaning they are capable of conducting both air-to-air and air- to-ground missions—few air-to-ground-capable airframes are regularly employed in this mission type.

3-3.   Most PLAAF combat aircraft date from the Cold War era. This status is changing rapidly, however, as newer and far more-capable types are being both purchased and indigenously produced. This creates a wide spectrum of capabilities within the PLAAF. Older aircraft are not capable of all-weather, around-the-clock operations, especially in ground-attack roles, and they employ mostly non-precision munitions with minimal targeting support. Newer aircraft employ precision-guided munitions, some with standoff capability, with targeting enabled by sophisticated onboard sensor suites. In general, more-capable aircraft are assigned to theaters with higher priority assets, while older aircraft are assigned to less important theaters. Nearly all PLAAF aircraft are either Russian-built, Russian-designed, or evolved from Russian designs. However, unique indigenous designs are starting to reach operational status. PLAAF combat aircraft capabilities are largely similar to those of their Russian equivalents, though the capabilities of the newest 5th-generation low- observable aircraft are not well known in the West.

3-4.   PLAAF maintenance and logistics capabilities are immature and under-resourced, but they are likely to improve rapidly in the near future. As part of the PLAAF's modernization efforts, maintenance and logistics were relegated to a secondary role behind acquisitions; this in turn had negative impacts on pilot training and aircraft readiness. In addition, the PLAAF had to modernize the relationship between its maintenance and logistics backbone and China's state-run industry. Historically, repair and maintenance was largely outsourced to industry, but this relationship has become increasingly unwieldy in the modern age. The PLA's new expeditionary requirements place further strain on limited maintenance and logistics resources. Solutions to these capability gaps are currently in the conceptual and testing phase, but have yet to be widely implemented.

3-5. PLAAF units are organized into brigades, comparable to USAF groups. Brigades are assigned to PLAAF bases, which are the grade equivalent of the U.S. Army's corps and are generally at the disposal of the theater commander. Some high-value PLAAF assets, however, are controlled by national-level headquarters, including some of the newest fighter divisions and the airborne corps.

3-6. PLAAF doctrine employs a tiered and layered approach to defending Chinese airspace. Ground-based air defenses in the form of medium- and long-range radar-guided missiles combine with manned aircraft to deny use of airspace to enemy platforms of all types. The PLAAF in general is far less concerned with seizing or maintaining air superiority or air dominance than U.S. forces—it views simply denying the enemy use of airspace to be adequate. PLAAF forces do not have significant capabilities outside of Chinese airspace: excepting a few island airbases off the Chinese coast, neither basing nor aerial refueling capabilities are adequate to project power any significant distance past the Chinese coastline and territorial waters. Expanding power projection across wider areas is a focus of PLAAF training and acquisition. It is likely that the PLAAF can successfully deny use of airspace to any opponent wishing to operate over Chinese territory or ocean areas covered by its surface-to-air missile umbrella.

3-7. PLAAF ground-attack capabilities are immature but improving. Around two decades ago, PLAAF attack aircraft were limited to daytime visual attacks using only unguided munitions. Today, the PLAAF operates hundreds of advanced 4th-generation attack aircraft capable of all-weather operations using precision munitions. The primary ground-attack missions of the PLAAF are strike and interdiction. Both of these missions attack targets some distance away from friendly troops, and both generally support the strategic and operational levels of war. PLAAF CAS capabilities are limited due to a lack of training and integration, but they are seen as critical to developing the combined arms brigade (CA-BDE) as a true combined arms force. Efforts are currently underway to advance CAS capabilities. Gaps in CAS operations include a lack of trained forward air controllers, a lack of air-ground communications, and a lack of developed tactics and guidelines for CAS employment.

3-8. The PLA's airborne corps is assigned to the PLAAF, and it is the PLA's strategic airborne unit. It comprises most of the PLA's Rapid Reaction Unit, a cohort of light, strategically mobile ground forces that can provide a significant military presence anywhere in China in a very short period of time. Though called a corps, the airborne corps mimics the group army in composition, though with fewer support assets. It consists of six maneuver brigades: four airborne infantry, one mechanized airborne infantry, one air assault, and one special operations forces, plus transport and support units. An airborne corps includes—

- Four airborne infantry brigades (three airborne infantry battalions and one towed light howitzer battalion per brigade).
- One airborne infantry (mechanized) brigade (three mechanized infantry battalions and one self-propelled howitzer battalion per brigade).
- One air assault brigade (three air assault battalions and one towed light howitzer battalion per brigade).
- One service support brigade.
- One special operations forces brigade.
- One air transport brigade (fixed- and rotary-wing transport capability).
- One headquarters unit.

(See figure 3-1 for a graphic depiction of an airborne corps.)

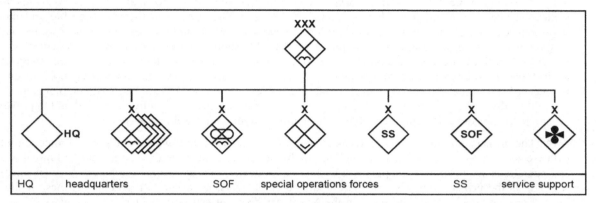

**Figure 3-1. Airborne corps (doctrinal)**

3-9.   The PLAAF has the air transport capacity to deploy either two light brigades simultaneously or half of the mechanized brigade. These units are employed in much the same way as the U.S. Army employs the 82nd Airborne Division: at least one brigade-size unit on very short-term deployment orders, with other units to follow if required.

# THE PEOPLE'S LIBERATION ARMY NAVY

3-10. China traditionally relegated its naval forces to a subordinate role behind its ground and strategic rocket forces. Much like the rest of the PLA, however, the People's Liberation Army Navy (PLAN) adapted and evolved along with the country throughout the 1990s and early 2000s. This evolution took the PLAN from being a littoral and riverine force to being a viable ocean-going force by 2009. The near-term goal for the PLAN is to evolve into a true "blue water" force by 2035. In practical terms, this means operating aircraft carriers, large surface combatants, a submarine force, a naval air wing, an amphibious assault force, and a logistics force across the Western Pacific and possibly into the Indian Ocean. The initial geographic limitation for this capability is out to the second island chain in the Western Pacific, which includes the Yellow Sea, the South and East China Seas, the Philippine Sea, and possibly the Sea of Japan. The PLAN has also expressed intent to develop a limited global capability, though this requires significant development of overseas basing support, long-range air and sea lift, and a global command and communication network that does not currently exist.

3-11. The PLAN is the world's second-largest navy by tonnage and the third-largest by number of major surface combatants. It also employs the world's second-largest submarine force, though the quality and modernity of its submarines varies significantly. The PLAN consists of five branches: the Surface Force, the Submarine Force, the Naval Air Force, the Coastal Defense Force, and the Marine Corps. These forces are deployed in three command navies, each with a specific geographic area of interest. The North Command Navy is responsible for the Yellow Sea, the East Command Navy is responsible for the East China Sea, and the South Command Navy is responsible for the South China Sea.

3-12. The Surface Force comprises the PLAN's surface combatant capability. This includes two aircraft carriers, around 20 modern, highly capable guided-missile destroyers, numerous lighter frigates and corvettes, missile boats, amphibious assault ships, and logistics vessels. The PLAN's aircraft carrier capability is still in its infancy, currently consisting of a single operational second-hand Soviet aircraft carrier, but an indigenously built one is scheduled to enter service in 2023. The PLAN intends to procure at least one more full-sized aircraft carrier, making its carrier fleet the second-largest and -most-capable in the world—behind only the United States. This timeline is aggressive, however, and is largely dependent on rapidly developing carrier air-wing tactics and techniques. The Chinese aircraft carriers operate navalized versions of 4 1/2th-generation Russian-derived multirole combat aircraft. Surface combatants are modern and capable, employing radar-guided missile-based integrated air defenses and a robust suite of antiship missiles of varying ranges. China has emphasized the importance of the antiship missile in its naval strategy for many decades, and much of its evolving navy is built around this capability.

3-13. The Submarine Force is the world's second-largest submarine fleet, although many boats are somewhat aged. Throughout the history of the PLAN, this force has been the most capable and most important sea-control capability. Although in recent years the Submarine Force has not been as well-funded as the Surface Force, it remains a key component of PLAN shore defense and antiaccess strategies. Most PLAN submarines are conventionally powered (diesel-electric) attack submarines that are employed against enemy shipping, emphasizing attacks on enemy surface combatants and troop ships. A small number of far more expensive, but far more capable, nuclear attack submarines complements the conventionally powered force. The PLAN's ballistic missile submarine capability is limited and unproven, though it appears development of a viable ballistic missile submarine fleet is proceeding.

3-14. The primary mission of the PLAN is defense of Chinese territorial waters and power projection into Chinese regional waters. This mission includes both operations against regional opponents in limited wars and operations against powerful opponents as a part of a theater-wide antiaccess campaign. As such, enemy shipping is the primary target set of PLAN surface combatants, naval aircraft, and submarines. While a limited ground-attack capability exists in the Naval Air Force, it can be considered a secondary mission, and far less important to PLA joint operations than U.S. Navy ground support missions are to U.S. joint operations. Chinese surface ships maintain a robust antiaircraft capability, and they are likely employed in concert with ground-based air defenses to deny use of wide geographic areas of airspace to enemy aircraft. The PLAN also maintains a robust Coastal Defense Force that employs shore-based antiship missiles and infantry to defend Chinese littoral waters and coastlines from amphibious assault or littoral naval operations.

3-15. The People's Liberation Army Navy Marine Corps (PLANMC) is the PLA's expeditionary amphibious warfare capability. Like the U.S. Marine Corps, it falls under administrative control of the navy, but it is equipped and organized in a manner similar to that of the army. Unlike the U.S. Marine Corps, however, the PLANMC does not have the PLA's heavy amphibious warfare mission—this belongs to the People's Liberation Army Army (PLAA). Instead, the PLANMC should be viewed as a light and strategically mobile force built to conduct expeditionary warfare missions away from Chinese shores.

3-16. The PLANMC consists of six maneuver brigades and associated command and support structure, totaling approximately 60,000-80,000 personnel. The PLANMC consists of a mix of mechanized and light forces organized specifically to support amphibious and littoral operations. PLANMC brigades are either assigned to the South Command Navy and focus their operations and training on the South China Sea, or they are held as national-level assets. A PLANMC brigade includes—

- Three combined arms battalions (CA-BNs).
- One self-propelled gun artillery battalion.
- One missile battalion (with man-portable air defense systems [MANPADS] and antitank guided missiles).
- One headquarters unit.
- One combat support battalion.
- One service support battalion.

(See figure 3-2 for a graphic depiction of a PLANMC brigade.)

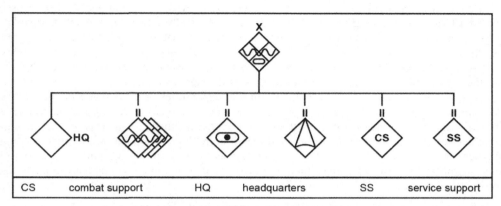

**Figure 3-2. PLANMC brigade (doctrinal)**

3-17. In addition to the PLAN, China operates both the China Coast Guard (CCG)—which falls under the control of the People's Armed Police (PAP)—and the Maritime Militia under the China Militia. The CCG operates a large number of smaller surface vessels and has responsibility for coastal security, patrol, antipiracy, antismuggling, search and rescue, and maritime police operations. The Maritime Militia operates an exceedingly large number of small surface vessels, most of which are converted or co-opted civilian fishing or transport ships. While both the CCG and Maritime Militia have very traditional assigned roles, China aggressively employs both as part of a hybrid warfare approach during the competition phase of conflict. CCG and Maritime Militia vessels regularly and aggressively operate in international waters—and sometimes in the territorial waters of China's neighbors—ostensibly to protect Chinese fishing and shipping operations. This approach enables China to influence activity throughout the Western Pacific without involving the large surface combatants of the PLAN. In a time of war, the CCG and Maritime Militia will likely take on roles similar to the PAP and China Militia in ground operations: performing low-intensity or mundane maritime tasks in order to free up more-capable ships for more demanding operations.

# THE PEOPLE'S LIBERATION ARMY ROCKET FORCE

3-18. The People's Liberation Army Rocket Force (PLARF) is the branch responsible for operating China's strategic missile and rocket forces and a substantial portion of the country's long-range tactical missiles. It is the largest missile force in the world, operating well over 1,000 short-range, medium-range, and intercontinental ballistic missiles (ICBMs) and over 300 long-range cruise missiles. There is not a U.S. equivalent to the PLARF. U.S. ICBMs are operated by the USAF, a very limited quantity of short- range ballistic missiles (SRBMs) are operated by the U.S. Army, and both the U.S. Navy and USAF operate cruise missiles.

3-19. The PLARF is descended from the Second Artillery Corps, China's original nuclear missile unit. It is a discrete military branch similar to PLAAF or PLAN. The PLARF operates using the same base-brigade-battalion structure as the PLAAF. PLARF bases are corps-grade, and each base operates multiple PLARF brigades. It is unclear exactly how the PLARF command interacts with theater commands (TCs), but it is known that each TC houses a PLARF contingent designed to integrate ballistic missile fires into its operations. PLARF missiles are considered strategic assets, and authority for their use is likely retained at the national level. It is also likely that high-value strategic missile assets, such as ICBMs and long-range ballistic missiles, are to be employed only by national authorities, while SRBM brigades may be attached to TCs in support of their operations.

3-20. The PLARF operates most of China's nuclear arsenal through a fleet of 60-70 ICBMs. China professes a "no-first-use" nuclear policy and maintains the nuclear missile fleet in a deterrent capacity—though there is an ongoing internal dialogue in China about possible situations were first-use may be necessary. The country at present does not possess an immediate second-strike nuclear capability, and it has relatively few nuclear warheads. Its nuclear strategy can be described as a minimal deterrence approach, possessing only the nuclear capability necessary to deter a nuclear attack. Future modernization accompanied by an expansion of the nuclear force, however, is a clear possibility.

3-21. In contrast to its relatively small nuclear force, the PLARF's conventional ballistic missile force is the world's largest and among the world's most technologically advanced and most capable. The PLA employs ballistic missiles as its primary precision deep-strike capability. Ballistic missiles target high-value assets, including air and seaports, supply depots, and command and communication nodes. Ballistic missiles represent a significant element of the Chinese antiaccess strategy. More-advanced ballistic missiles are designed specifically to engage hardened or mobile high-value assets, such as aircraft carriers and antiballistic missile systems. Chinese ballistic missile capabilities represent one of the PLA's strongest investments in system warfare, as ballistic missiles asymmetrically destroy or neutralize assets that traditionally required force-on-force methods to effectively attack. China is not a signatory to the Intermediate-Range Nuclear Forces Treaty, and thus it is free to develop short- and medium-range missiles of all types. Due to the country's strict no-first-use nuclear policy and its substantial investment in ballistic missile capabilities, it is unlikely that China will ever voluntarily downgrade its conventional missile-strike capability.

3-22. While many PLARF ballistic missiles are strategic assets, hundreds of SRBMs may be used to target high-value tactical assets or in support of tactical-level missions. PLARF ballistic-missile launchers are mobile, and their facilities hardened, making it difficult to target or neutralize these systems prior to launch. The PLARF also operates much of China's land-based cruise-missile fleet. Unlike the United States, PLA cruise missiles are seen as complementary to the ballistic missile force, rather than the other way around.

3-23. PLARF missile tactics are advanced and mature, taking full advantage of the suite of capabilities available. PLARF units will work along with other national assets, such as cyber warfare, EW, and air and naval forces, to neutralize or destroy antiballistic missile and air defense systems using a system warfare, combined-arms approach. PLARF ballistic missiles are accurate and precise, integrated with advanced surveillance and targeting capabilities, equipped with advanced penetration aids and countermeasures, and can be employed in sophisticated structured attacks. The PLA will seek to deplete expensive and rare antiballistic interceptors; defeat or destroy air defense radars; and either sink, damage, or threaten antiair surface naval vessels, giving it freedom of maneuver when employing its missile force.

# THE PEOPLE'S LIBERATION ARMY STRATEGIC SUPPORT FORCE

3-24. In 2016, the PLA established the People's Liberation Army Strategic Support Force (PLASSF). The PLASSF is a unified command with responsibility for national-level space, cyber, and EW missions, along with national-level intelligence support to military operations and information operations. There is not a U.S. equivalent to the PLASSF. It mixes capabilities taken from numerous military and nonmilitary government agencies under a single command, much of which is controlled directly by the Central Military Commission (CMC).

3-25. The PLASSF represents the evolution of several basic Chinese warfighting principles into a unified command structure. It combines numerous capabilities along the competition continuum and seeks to employ them in a more structured and coherent way than the largely piecemeal approach previously used by the PLA. The PLASSF emphasizes system warfare as its underlying operational principle, seeking to find asymmetric approaches to neutralizing or otherwise offsetting highly capable enemy systems.

3-26. The PLASSF was likely developed to help reduce the PLA's acknowledged shortcomings in joint operations, particularly joint intelligence collection and distribution. It echoes the general PLA trend to centralize capabilities that are deemed insufficient or immature, with the intent of managing them more effectively at a national level. The PLASSF also supports the PLA's initiative toward Military-Civil Fusion. The intent is to enhance cooperation between the PLA, nonmilitary organizations, and industry. This initiative is most clear when it comes to space capabilities: the PLASSF will likely work very closely with the China National Space Administration and the country's space industry partners to develop dual-use space capabilities. The advantage to this approach is that it allows military space development while being minimally provocative internationally, as all initiatives have a legitimate civilian purpose.

3-27. The other major change driving the PLASSF is the integration of national intelligence. China's intelligence apparatus is enormous, complex, and almost completely opaque to outsiders. It consists of numerous and occasionally competing intelligence organizations that focus on both domestic and international intelligence gathering. The PLASSF integrates electronic intelligence (ELINT), signals intelligence (SIGINT), long-range surveillance, and information operations capabilities under a single command. China believes this greater centralization will result in better use of national intelligence assets while expediting critical intelligence information—such as targeting data—to commanders and weapons systems that need it.

3-28. The PLASSF likely used the PLARF as the blueprint for its organization. National-level assets are retained at the top level of command, while lower-level assets are allocated to TCs as missions require. PLASSF intelligence operations are split into strategic intelligence and tactical intelligence. Strategic intelligence consists of collection efforts focused on long-term issues of national importance. Tactical intelligence is composed of collection efforts that directly support PLA operations, such as ground intelligence, air- and space-based surveillance, deep ELINT, and tactical-level information operations

This page intentionally left blank.

# Chapter 4
# Tactical System Warfare

This chapter provides an overview of system warfare as it is applied at tactical echelons. System warfare is the basic theoretical construct the People's Liberation Army (PLA) uses to build capabilities at all echelons. People's Liberation Army Army (PLAA) units historically did not often task-organize, especially at tactical echelons; as such, the system warfare concept is relatively immature at these levels. System warfare requires a modular approach and substantial task organization in order to be successful. It also requires enhanced cooperation between sister and adjacent units, a skill that is similarly underdeveloped in PLAA formations. Nonetheless, it appears that the PLAA is deeply committed to the system warfare methodology, and it will likely govern PLAA tactical operations for the foreseeable future.

## TACTICAL SYSTEM WARFARE PRINCIPLES

4-1.   Land warfare traditionally involves imposing one's will upon an uncooperative opponent by the use of direct lethal attack or threat of attack. Maneuver is the process by which one puts oneself in an advantageous position to conduct an attack. Thus, ground combat components traditionally maneuver with the objective of physically isolating their opponents, then defeating them in detail; creating numerical or firepower advantages at key positions on the battlefield; or deceiving opponents into believing their position to be hopeless or otherwise indefensible. The PLAA still views land warfare through this lens, but it sees attacks that do not directly target one's opponent with the threat of physical danger as equal to or more important than direct lethal attacks on the modern battlefield. For example, an opponent may be isolated by disabling network or communications connectivity, rather than by being physically surrounded and cut off. An opponent may believe its position untenable by having vulnerable nodes of its systems rendered ineffective, rather than having the entire system destroyed. An opponent may view continued resistance as futile—not because of the direct threat of physical force—but because it has been deceived into thinking its situation is hopeless. In other words, system warfare takes the basic principles outlined by Sun Tzu and Mao and applies them to conflict utilizing modern, high-technology weapons systems that are manned and operated by professional soldiers.

4-2.   The PLA describes system warfare using several different official names, including system destruction warfare and system confrontation. All of these terms refer to the same basic concept: a type of conflict wherein systems clash with one another in an attempt to neutralize, destroy, or offset key capabilities and thus grant one side a decisive advantage. System warfare differs from traditional Western military thinking in that it does not necessarily consider the human doing the fighting as the most important element of a combat system. Humans are viewed as subcomponents of a system of systems, to be assessed and targeted much as any other subcomponent. System warfare represents Military Thought supporting the Defense Theory and Defense Doctrine concepts within People's War in Conditions of Informationization. (See chapter 1 for more information on Defense Theory and Defense Doctrine.) The backbone of the system of systems is the network; the importance of this backbone in-turn gave rise to the concept of network-centric warfare.

4-3.   The system warfare concept consists of two basic ideas: creating purpose-built operational systems that combine key capabilities under a single command, and the use of these operational systems to asymmetrically target and exploit vulnerable components of an opponent's system. If done effectively, this method will render the opponent's key systems ineffective or otherwise unable to function. The PLA believes that by effectively destroying, isolating, neutralizing, or offsetting key capabilities, the enemy's will or ability to resist will degrade and victory will be achieved. Building operational systems is similar to creating task forces, but broader in scope, attempting to create a comprehensive suite of capabilities under a single command. An operational system consists of a number of groups—subordinate entities that are custom-built for a specific mission, task, or purpose. Virtually every battlefield function is represented by one or more groups; this publication discusses only the most significant groups as employed by tactical-level units.

4-4.    At the tactical level, operational groups are also referred to as combat teams or combat groups. While the PLAA stresses a modular approach to building operational systems, it also recognizes that the less radical the reorganization, the more cohesive a unit will be. The combined arms battalion (CA-BN) structure is designed to reflect this. It is the basic building block of the tactical operational system, and it is intended to be employed in something close to its organic or peacetime configuration. Conversely, the combined arms brigade (CA-BDE) is intended to be easily augmented or task-organized as conditions dictate, flexibly employing a variety of subordinate CA-BNs, supporting battalions, or other nonorganic capabilities.

# PLANNING AND ORGANIZING PLAA OPERATIONS

4-5.    The PLAA places great emphasis on planning. A meticulous approach to operations underpins the PLAA throughout its history, and this tradition remains in place today. Though the PLAA seeks to gradually move to an increasingly decentralized leadership philosophy, careful planning at all echelons will remain a basic principle.

4-6.    Movement toward greater decentralization, modular combined arms units, widespread downsizing, and the creation of new headquarters at the national and theater command levels have significantly complicated the PLAA's employment of command and support relationships. The PLAA was traditionally a strongly centralized and hierarchical force, relying on a mixture of discipline and obedience to overcome shortages in combat technology and firepower. The evolved PLAA recognizes that greater decentralization requires a more sophisticated understanding of command and support relationships; improved professional military education for commanders, staffs, and noncommissioned officers; continued PLAA doctrinal adjustments as new equipment and technologies enter the force; and a task-oriented approach that underpins the challenging requirement of building operational systems.

## PLAA PLANNING PROCESS

4-7.    The PLAA planning process is broadly similar to the U.S. Army's military decision-making process. The primary outputs of the planning process are the objectives and scheme of maneuver, along with the structure of the operational system being used to conduct the operation. The operational system may include multiple subordinate operational systems in addition to specialized supporting systems.

4-8.    The PLAA planning process consists of five steps, each with distinct outputs. Subordinate units conduct planning in sequence with their higher echelon headquarters and adjust their plan according to inputs received from it. The planning process seeks to achieve unity of purpose throughout the operation, ensuring that all subordinate commanders know their role, mission, and place in the wider operation. Figure 4-1 on page 4-3 depicts the PLAA planning process.

### Step 1: Assess the Situation

4-9.    During this step, the commander assesses four critical components of the upcoming confrontation: terrain, enemy forces, friendly forces, and other considerations, such as civilian presence and political elements. The outputs from this step include an early outline of the desired friendly operational systems, a comprehensive report on enemy strength and disposition enabled by reconnaissance and intelligence support, and a thorough analysis of all other factors in and around the battlefield that may influence operations for either side.

## Step 2: Make Decisions

4-10. This step requires commanders to make several key decisions that determine the direction of the rest of the planning process and the operation at large. Commanders must establish the purpose and objective of the operation, the general scheme of maneuver to be employed to achieve this purpose and objective, the basic structure of the operational systems under their command, and key offensive and defensive points of interest. The output of this step allows subordinate commanders to begin their own planning process, staffs to begin building operational systems, and command posts to deploy in order to establish command systems. Reconnaissance operations should also commence at this point in order to support the intelligence requirements for the wider operation.

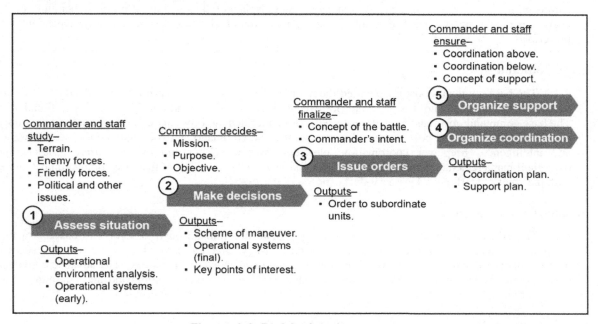

**Figure 4-1. PLAA planning process**

## Step 3: Issue Orders

4-11. This step consists of two main phases. First, commanders conceptualize the battle to their staffs, accounting for enemy strengths and disposition, friendly forces, and the higher echelon commanders' intent. The staff then creates an order that clearly lays out the commander's concept in such a way that subordinate units can easily understand it. Second, the order is issued to subordinate units, either verbally or in writing. Orders are ideally issued to subordinate commanders in groups, allowing subordinates to ask questions and collaborate with one another prior to conducting their own planning process.

## Step 4: Organize Coordination

4-12. This step refines the orders issued in step 3. Staffs conduct planning that synchronizes purpose between subordinate units, ensures that sister units' schemes of maneuver are integrated, provides useful predictions about how the battle will be conducted, clarifies issues of confusion with subordinate commanders, and ensures that shared resources and enablers—such as communications and network support—are deconflicted.

## Step 5: Organize Support

4-13. This step organizes external capabilities that support the operation. Examples include combat support, logistics support, equipment support, and political support. Specific capabilities include firepower and artillery support, information warfare, sustainment, casualty evacuation, and medical support. This step requires staffs to interact with external enablers; higher echelon staffs may provide assistance in building this part of the plan. Commanders must account for available resources, mission requirements, and friendly forces when developing the support plan.

## PLAA COMMAND POST OPERATIONS

4-14. The PLAA defines a command post as a temporary command structure formed around a commander and associated staff. Up to four command posts are established in support of an operational system: a base command post, an advance command post, a rear command post, and a reserve command post. Command posts are led by a commander and manned by various command groups. Certain command posts—most likely those more forward on the battlefield—may be designed as mobile command posts, able to conduct command post operations while rapidly emplacing and displacing, thus making it more challenging for the enemy to detect and target. The number and type of command posts are situationally dependent, not prescribed.

### Base Command Post

4-15. The base command post (also called the main command post) is the primary command structure for the operational system, and it is the center for executing combat command throughout the operational area. The base command post must be well protected, well concealed, and roughly central within the operational area. The base command post houses all command groups: command and communication, reconnaissance and intelligence, firepower coordination, electronic warfare (EW) and cyber warfare, engineering, battlefield management, and political work. The commander operates out of the base command post, along with the chief of staff and all primary staff officers.

### Advance Command Post

4-16. The advance command post (also called the forward command post) is a forward-based structure designed to enhance command and communication in a key direction of the battle. Ideally, it is situated near the main offensive or defensive effort, and it is well concealed and protected. The advance command post is typically led by the deputy commander and staff, and it houses a command and communication group, a reconnaissance and intelligence group, and a firepower coordination group.

### Rear Command Post

4-17. The rear command post (or alternate command post) is responsible for logistics and equipment support, along with rear area security. It consists of a combat support and service support group headquarters, a political work group, and security personnel. The commander of the combat support or service support group is likely the commander of the rear command post. The rear command post serves as the primary backup command post in case the base command post is destroyed, neutralized, or otherwise compromised. Depending on the situation, the deputy commander may choose to operate from the rear command post.

### Reserve Command Post

4-18. The reserve command post is a backup structure in case one of the other three command posts is compromised, damaged, destroyed, or otherwise neutralized. It is necessarily a smaller and less-capable structure than the base command post, but it must be able to conduct all base command post operations. The reserve command post may also serve as an interim command post in case more forward command posts are temporarily unavailable due to movement or enemy action.

## PLAA CONTROL MEASURES

4-19. The PLAA employs zones to define the combat area—a designated area assigned to a military, paramilitary, or security force in which combat or security operations occur—for both offensive and defensive actions. These are made up of several secondary zones or areas; each has specific characteristics and is designed to be occupied by one or more groups. Control measures are established for each echelon and are typically nested within the combat area for the higher echelon unit. Security zones are established throughout a zone any time security operations are conducted. Annihilation zones—specific geographic areas to which the enemy is to be lured or driven and then destroyed—may also be created in any of the secondary zones.

## Offensive Control Measures

4-20. The offensive zone typically consists of four secondary zones or areas: the deep area, the frontline zone, the reserve zone, and the garrison zone. These control measures are described in paragraphs 4-21 through 4-24. (See figure 4-2 on page 4-6 for an illustration of offensive control measures.)

### Deep Area

4-21. The deep area is the territory past which a unit's organic sensors and weapons can operate. For a CA-BDE, this typically means the area past which its rocket artillery and targeting support can operate. Units operating in deep areas are usually more autonomous and can expect minimal support from their parent unit.

### Frontline Zone

4-22. The frontline zone is the territory in which the main offensive action is to take place. Typically, first-line objectives and the enemy's main defensive position are located in the frontline zone. The entire frontline zone should be within the range of the offensive group's fire support. The frontline zone typically contains a security zone on its forward edge, where security, reconnaissance, and counterreconnaissance activities take place.

### Reserve Zone

4-23. The reserve zone is the territory just to the rear of the frontline zone that typically houses depth attack groups, command groups, firepower groups, and forward logistics bases (For more information on PLAA groups, see paragraphs 4-32 through 4-60.) The reserve zone also serves as a defensive bastion against enemy counterattacks and as a secure location through which follow-on forces and supplies can move into frontline and deep areas.

### Garrison Zone

4-24. The garrison zone consists of rear areas not actively occupied by the offensive group. Supporting capabilities such as logistics, EW, and long-range artillery reside here. Garrison zones typically contain one or more security zones that surround key positions such as bases, supply routes, or command posts. The People's Armed Police (PAP) may take on much of the security load in garrison zones in order to free up PLAA forces for more-intense duties.

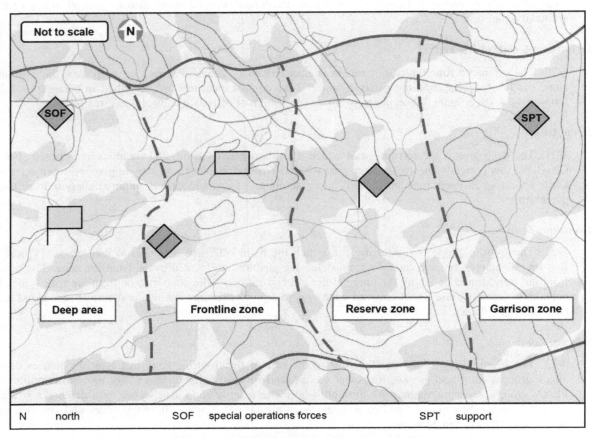

**Figure 4-2. Offensive zone (conceptual)**

## Defensive Control Measures

4-25. The defensive zone typically consists of five secondary zones or areas: the deep zone, the frontal blocking zone, the frontier defense zone, the depth defense zone, and the rear defense zone. These defensive control measures are described in paragraphs 4-25 through 4-29. (See figure 4-3 for an illustration of defensive control measures.).

### Deep Area

4-26. The deep area is defined in the same way as the deep area in an offensive zone. Usually it is the area in which independent units conduct reconnaissance and disruption activities prior to heavy enemy contact.

### Frontal Blocking Zone

4-27. The frontal blocking zone is the forward-most area occupied by the defensive group. Screening, reconnaissance, and counterreconnaissance missions are the primary focus in this zone. Groups in the frontal blocking zone are usually used to disrupt, canalize, and slow down enemy assault forces in preparation for a future decisive counterattack.

### Frontier Defense Zone

4-28. The frontier defense zone is the zone that typically contains the main line of defense, consisting primarily of the frontier defense group. The bulk of the defensive group is deployed here, making best use of terrain to slow, disrupt, and degrade enemy units as they conduct their attack. Most fortifications and other defensive engineering works are focused in the frontier defense zone.

*Depth Defense Zone*

4-29. The <u>depth defense zone</u> is the zone that houses the depth defense group, and it may contain the firepower strike or coordination group or parts of the command group. The depth defense zone is designed to facilitate the commitment of the depth defense group to defeat the enemy's main effort. (For more information on depth defense groups, see paragraph 4-49. Firepower strike groups are described in paragraph 4-51, and firepower coordination groups are described in 4-36.)

*Rear Defense Zone*

4-30. The <u>rear defense zone</u> is the zone that houses the combat reserve group, long-range firepower, command posts, and forward logistics support. A secondary defensive line is often established in the rear defense zone as a fallback position in case forward zones are overrun.

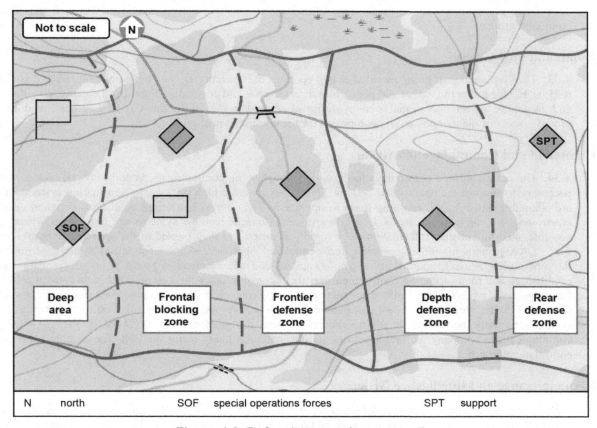

**Figure 4-3. Defensive zone (conceptual)**

# TACTICAL SYSTEM WARFARE METHODOLOGY

4-31. Building the operational system is one of the primary outputs of the PLAA's planning process. Commanders are given general guidelines on how to assemble the groups that comprise the operational system, and then they build the operational system to suit the mission and situation. Using the commander's purpose, intended scheme of maneuver, and analysis of enemy forces and other battlefield conditions, groups are formed using the available force pool, weighting specific groups with the appropriate levels of firepower, mobility, and protection. In tactical ground operational systems, groups can be organized into five general categories: command groups, offensive groups, defensive groups, firepower strike groups, and support groups. Not all operational systems will have all groups; most will consist of command groups supported by one or more additional groups.

> *Note.* PLA literature uses a very wide variety of names for these groups; its approach appears to be unstandardized and still somewhat in development. This list represents a cross-section of different group names and mission descriptions and is not exhaustive.

## COMMAND GROUPS

4-32. Command groups are the components of the operational system that enable leadership to control the system, make rapid and effective command decisions, inform subordinate commanders, and coordinate capabilities. They are built using a variety of different force pools—from command posts and staffs, to reconnaissance and intelligence units, to network support units. Some functions, such as engineer and chemical support, are prescribed in both the command group and a support group; it is unclear how the decision to place these units is made. The following paragraphs describe several of the most important command groups.

### Command Post Group

4-33. The command post group establishes and operates the operational system's command posts. It is also responsible for protecting command posts from direct assault and concealing them from enemy surveillance and targeting. In an operational system designed around a CA-BDE, the CA-BDE's command post provides most of the manpower and equipment for the command post group.

### Command and Communication Group

4-34. The command and communication group establishes the communications and network architecture necessary to support the operational system. The PLAA places a high priority on information systems that are adaptable and reliable, enabling them to operate in a variety of battlefield conditions or when under electromagnetic or network attack by enemy forces. Communications systems are automated wherever possible, enabling critical information to be rapidly distilled and then passed to the appropriate consumer. The PLAA prefers to use a top-down and centralized approach to its information systems, ensuring interoperability across disparate units. The command and communication group is also responsible for establishing and monitoring the control measures that control information systems within its combat area—functions such as assigning radio frequencies, deconflicting electronic emitters, and ensuring network security. Finally, the command and communication group is responsible for protection of the operational system's network backbone. In an operational system designed around a CA-BDE, the CA-BDE's communications company within the service support battalion likely forms the basis for the command and communication group.

### Reconnaissance and Intelligence Group

4-35. The reconnaissance and intelligence group is comprised of reconnaissance forces and the operational system's intelligence personnel, task-organized under a single group structure. This structure enables close integration of intelligence and reconnaissance, outlined in detail in chapter 6. The reconnaissance and intelligence group is responsible for developing the reconnaissance plan, conducting reconnaissance in support of the operational system, and disseminating relevant intelligence to all members of the operational system. In an operational system designed around a CA-BDE, the CA-BDE's reconnaissance battalion and intelligence headquarters section form the core of the reconnaissance and intelligence group.

### Firepower Coordination Group

4-36. The firepower coordination group, also called the firepower coordination center, is responsible for integrating firepower capabilities of all types into a coherent, synchronized plan in support of the operational system. Firepower capabilities include, but are not limited to, tube and rocket artillery, ballistic and cruise missiles, fixed- and rotary-wing aircraft, network, electromagnetic, and information attack. The firepower coordination group is enabled by the firepower coordination system, an automated or semiautomated command and communication network that distributes targeting and system data across a variety of integrated firepower systems and seeks to streamline targeting and fire control. The firepower coordination group is closely linked with all of the firepower groups.

### Electronic and Network Warfare Group

4-37. The electronic and network warfare group is comprised of electronic countermeasure forces, network warfare forces, and other associated personnel. Its missions include electronic reconnaissance, interference, sabotage, targeting, network defense, and cyber intrusions against vulnerable enemy networks.

### Battlefield Management Group

4-38. The battlefield management group is responsible for managing personnel and administrative functions in support of the operational system. It consists of the personnel staff officer, supporting staff, and support teams.

### Political Work Group

4-39. The political work group is responsible for political support to the operational system. This includes managing the political messaging of the commander, ensuring synergy of purpose for all troops in the operational system, conducting propaganda operations in support of friendly operations and against enemy operations, and supporting psychological warfare operations. The political work group is led by the political officer and manned by supporting staff.

## OFFENSIVE GROUPS

4-40. Offensive groups are the elements of the operational system that are responsible for fixing, assaulting, and annihilating enemy forces. Offensive groups are deployed in such a way as to ensure substantial numerical and firepower advantage over the enemy: PLAA guidelines suggest offensive groups seek a four-to-one advantage in maneuver forces, a five-to-one to seven-to-one advantage in artillery firepower, and three PLAA antitank systems for each anticipated enemy armor system. An offensive group deploys in a focused manner, establishing the depth and three-dimensional organization necessary to overwhelm the enemy when an assault commences. Offensive groups are integrated with firepower strike groups in order to achieve synchronization between maneuver and fires capabilities. Offensive tactics and techniques are discussed in greater detail in chapter 7.

### Advance Group

4-41. The advance group is composed of troops that take advanced positions ahead of a primary attack, conducting a mission roughly similar to the advance guard. The group provides security for the main body, screens against enemy troops, conducts counterreconnaissance, occupies terrain favorable for the main body, and initiates contact with the enemy main body. After completing these tasks, the advance group typically assumes a security or reserve role.

### Frontline Attack Group

4-42. The frontline attack group consists of those troops whose mission is to conduct the initial assault against a hardened enemy target, with the goal of achieving a breach or other small penetration. The frontline attack group includes infantry units, armor units, firepower units—likely in the form of mortars supported by artillery—and air defense, antitank, engineer, chemical weapons defense, and electronic countermeasure units. The group's primary task is to break through and capture enemy first-line positions. The frontline attack group is typically subdivided into assault teams and firepower teams, and it seeks to support each assault team with a firepower team. The frontline attack group attempts to concentrate combat power on the narrowest possible front, then launch a well-supported attack. Armored units likely form the core of this main thrust, supported by infantry. Once the assault is successful, the frontline attack group transitions to a security-or vigilance-role, providing flank and rear security for the depth attack and thrust maneuvering groups as they commence their follow-on missions.

### Depth Attack Group

4-43. The depth attack group (or in-depth attack group) consists of those troops that advance deep into the enemy position once the initial breach is achieved. The depth attack group likely consists of the best-available armored forces, supported by mechanized infantry, air defense, antitank, engineer, chemical weapon defense, and electronic countermeasure units. The tasks of the depth attack group include breaking into the depth of the enemy's defense, seizing critical terrain or other targets deep in enemy territory, annihilating the enemy's in-depth positions, and occupying positions favorable for defense against the enemy's counterattack. The depth attack group is staged in order to ensure that the penetration does not lose momentum as forces tire, face challenging terrain, or meet enemy resistance.

### Thrust Maneuvering Group

4-44. The thrust maneuvering group consists of those troops that exploit the advantages created by the depth attack group. The thrust maneuvering group may be a highly mobile armored force or an air assault force, enabled by light artillery, antitank, air defense, engineer, and EW units. The thrust maneuvering group continues the attack against enemy deep positions, targeting command nodes, supply areas, and key terrain. Its most important missions, however, are to cut off enemy retrograde routes—enabling the annihilation of enemy units—and to spoil or disrupt counterattacks as the enemy attempts to reinforce its units that are caught in the annihilation zone(s).

### Combat Reserve Group

4-45. The combat reserve group consists of those troops that remain in the operational system's rear area, with missions to reinforce the frontline attack group or depth attack group if necessary, spoil enemy counterattacks that threaten other offensive groups or the rear area, and deal with any other situations that may arise in the operational system's combat area. The combat reserve group likely consists of either recently used armored units on a rest cycle or lighter mobile units that are able to respond rapidly to developing situations. The combat reserve group is augmented by antitank and engineer units.

### DEFENSIVE GROUPS

4-46. Defensive groups are those groups charged with defending friendly forces, systems, or key terrain from enemy attack. Defensive groups conduct either a position-based or mobile defense, with the intent of blunting the enemy's attack, attriting enemy forces, and enabling friendly counterattack and transition to the offense. Defensive groups are integrated with firepower strike groups in order to enhance their combat power and defeat superior enemy forces. Defensive tactics are discussed in greater detail in chapter 8.

### Cover Group

4-47. The cover group is made of those troops assigned to a cover mission in support of the operational system's main body. The group's primary tasks are to conduct counterreconnaissance, defend stubbornly in the face of an enemy attack, screen the main body, and cover the main defensive line's deployment and disposition. After the cover mission is complete, the cover group may withdraw to deeper areas and conduct vigilance or security operations, serve as the combat reserve group, or continue to operate as part of the main defensive line. If possible, a component of the cover group will attempt to remain behind the enemy's main advance in order to conduct operations behind enemy lines. The cover group consists of reconnaissance or light armored units enabled by light artillery, antitank, antiair, and EW units. Additional information on the cover mission can be found in chapter 6.

### Frontier Defense Group

4-48. The frontier defense group consists of those troops assigned to the main line of defense. Frontier defense groups are further divided into a main defensive direction group and secondary defensive direction groups, depending on the enemy's predicted course of action. The main defensive direction group is weighted more heavily with numbers and firepower; secondary defensive direction groups are accordingly less powerful, proportional to the commander's assessment of the enemy's approach. The primary tasks of the frontier defense group are to hold the defensive line, blunt the enemy's offensive attack, inflict heavy casualties upon the enemy, buy the commander decision space, and inform the commander about how best to commit reserves and counterattack forces. In a position-based defense, the frontier defense group is likely much larger, as a position-based defense requires the defender to hold a fortified position with minimal maneuver. The frontier defense group consists of infantry and armor, augmented by artillery, antitank, engineer, anti-chemical weapon, and electronic countermeasure units. It likely consists of a higher proportion of infantry compared to other defensive groups.

## Depth Defense Group

4-49. The depth defense group consists of those troops assigned to defend deep areas of the defense. Its primary mission is to conduct counterattacks against enemy penetrations; reinforce weak areas of the frontier defense group's lines; defeat aerial incursions into the rear defense area; and encircle, isolate, and assault any enemy forces operating in rear areas. The depth defense group must be mobile, and it likely contains a high proportion of armored and mechanized forces. In a mobile defensive operation, the depth defense group is the decisive component. Commitment of the depth defense group is the decisive point in a defensive battle. The depth defense group is augmented by artillery and antitank units.

## Combat Reserve Group

4-50. The combat reserve group is composed of those troops that are retained in depth areas, with a mission to maintain security and reinforce the main defensive line, if necessary. The group may also conduct counterattacks against enemy penetrations. The combat reserve group likely consists of light, mobile troops or troops on a rest cycle following previous action.

## Firepower Strike Groups

4-51. Also called fire assault groups, these groups provide the bulk of the operational system's firepower. Firepower strike groups include a variety of heavy weapons, such as artillery systems, mortar systems, air defense systems, and antitank systems, and they include electromagnetic attack and psychological attack systems. These disparate systems are integrated into the operational system by the firepower coordination group, and they can further be subdivided into firepower teams charged with supporting small subordinate elements of the operational system.

## Artillery Group

4-52. The artillery group consists of the indirect fire elements of the firepower strike group. These include 122-mm and 155-mm tube artillery systems, light and heavy multiple launch rocket systems, and short-range ballistic missile systems. The PLAA's firepower is seen as the backbone of its operations, and the artillery group provides most of the tactical-level firepower to the operational system commander. The artillery group is positioned in deep areas, and its primary missions are to—

- Support scouting and security operations.
- Disrupt or pre-empt advancing or unfolding (pre-attack positioning) enemy troops.
- Support defensive groups.
- Disrupt or defeat enemy flanking and encirclement attempts.
- Disrupt enemy penetrations and counterattacks.
- Suppress enemy artillery.
- Strike at key targets, such as enemy command posts and supply areas.

Broadly speaking, fire support missions are conducted by tube artillery units, while annihilation, suppression, and counterfire missions are conducted by rocket artillery units. Missile units, if available, conduct precision deep strike missions. The artillery group is built around artillery battalions combined in such a way as to achieve a combined arms effect.

## Air Firepower Strike Group

4-53. The air firepower strike group (also called the aerial firepower assault team) consists of the operational system's rotary-wing airframes that are capable of ground-attack operations. The PLAA employs its helicopters as airborne fighting vehicles, closely attaching them to ground units and seeking to maximize their unique capabilities against vulnerable parts of enemy formations. The air firepower strike group is employed in a variety of ways: as a highly mobile reserve, attacking enemy penetrations and supporting friendly counterattacks; as mobile artillery, targeting enemy artillery systems and other weapons that were not destroyed by the artillery group; as an antitank unit targeting enemy tanks and other armored vehicles; as a deep penetration force targeting enemy command, EW, and sustainment systems; as reconnaissance and surveillance platforms; and as security and vigilance platforms.

### Antitank Group

4-54. The antitank group consists of those troops specially equipped and tasked with targeting and destroying enemy armored vehicles. The antitank group consists of a combination of mounted antitank guided missile launchers, dismounted antitank guided missile teams, and antitank artillery—mainly in the form of assault guns. The antitank group is positioned near the main direction of attack or defense, and it is employed as a rapid, mobile reserve whenever enemy armored forces are encountered. If armored targets are not available, the antitank group can serve as one or more firepower teams, augmenting assault teams as they carry out their missions. The PLAA employs a high density of antitank weapons throughout its formations, and employment of antitank groups is a key planning consideration.

### Mobile Artillery Group

4-55. The mobile artillery group is comprised of highly mobile artillery and mortar systems that provide direct, close-in fire support to offensive and defensive groups of all types. Missions for mobile artillery groups include suppressing or destroying enemy positions, delivering obscurants or illumination, conducting short-range counterfires, and disrupting enemy reinforcement or counterattack maneuvers. The mobile artillery group is viewed as vulnerable to counterfire and direct fire, and rapid movement is considered key to its survival. The mobile artillery group is likely built around either 82-mm or 120-mm rapid-fire mortars or mobile 122-mm self-propelled howitzers.

### Air Defense Group

4-56. The air defense group is comprised of most of the operational system's air defense weapons systems, plus associated sensors and command nodes. At the tactical level, this consists of short-range air defense (SHORAD) and man-portable air defense systems (MANPADS), plus a variety of towed and self-propelled antiaircraft guns (SPAAGs). The air defense group's primary mission is to deny the use of airspace over the battlefield by enemy aircraft through a mixture of deterrence and lethal engagements of in-range aircraft. The air defense group may also conduct air ambushes, intended to surprise and destroy enemy aircraft. Per Chinese doctrine, the group should be employed using a defense-in-depth approach, with deployments weighted according to the value of the assets defended.

### SUPPORT GROUPS

4-57. Support groups are responsible for the operational system's support and sustainment functions and rear area security functions. At the CA-BDE level, these are built primarily using the CA-BDE's two support battalions: combat support and service support.

### Combat Support Group

4-58. The combat support group consists of a variety of combat support capabilities. At tactical echelons, most of this group is comprised of the CA-BDE's combat support battalion. Capabilities include engineering, mobility, protection, anti-chemical weapons, and communications. Special emphasis is given to obstacles: there are discrete barrier set-up and removal teams. The combat support group also provides deception and camouflage support and smokescreen generation.

### Rear Area Support Group

4-59. The rear area support group is comprised of most of the operational system's supply and logistics capabilities. The rear area support group is divided into two subgroups: logistics support and equipment support. The logistics support group conducts the supply mission, while the equipment support group performs maintenance.

### Psychological Warfare Group

4-60. The psychological warfare group is comprised of specially trained soldiers that provide support to frontline units. The group's mission consists of three tasks: present a clear picture of the enemy's psychological situation, conduct psychological attacks on the enemy in support of the operational system's operations, and protect friendly troops from enemy psychological attack. The psychological warfare group also works carefully with the political group to build consensus and morale among friendly troops.

## TACTICAL-LEVEL NOTIONAL OPERATIONAL SYSTEM

4-61. This section describes one possible construction of a tactical-level operational system, also called a combat group. Since operational systems are, by definition, ad hoc and task oriented, there is not one single or correct way to assemble them. Each operational system is different due to battlefield characteristics, the nature of the enemy, and the preferences of the commander. This section is intended to demonstrate the building-block nature of the operational system and how a system of systems is created.

4-62. The example combat group is assembled by a group army, using a medium CA-BDE and a heavy CA-BDE as the two main force providers. The group army's artillery, air defense, aviation, and special operations forces (SOF) brigades provide additional forces to the combat group. The group is assembled in order to conduct an attack against a fortified enemy. The group army command and staff build a single, coherent combat group, minimizing ad hoc or unnecessary reorganization and maintaining the cohesion of subordinate battalions wherever possible.

4-63. The force pool that the combat group is built from consists of all the organic units of the two CA-BDEs, plus augmentation from the group army. Available forces to build the combat group, with units in italics being reinforcements from the group army, include—

- Four medium CA-BNs.
- Four heavy CA-BNs.
- Two reconnaissance battalions.
- *One unmanned aircraft system (UAS) reconnaissance section.*
- *One SOF company.*
- *One scout helicopter company.*
- One UAS targeting section.
- Two artillery battalions.
- *One heavy multiple rocket launcher (MRL) battery.*
- *One light MRL battery.*
- *One attack helicopter company.*
- *One medium lift helicopter company.*
- Two air defense artillery battalions, consisting of three SPAAG batteries and one SHORAD battery each.
- *One heavy SPAAG or missile battery.*
- *One medium-range surface-to-air missile (SAM) battery.*
- *One electronic air defense battery.*

(See figure 4-4 on page 4-14 for a graphic depiction of an operational system force pool.)

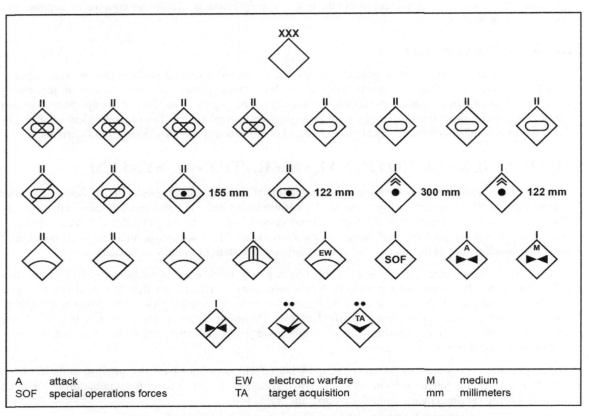

| A | attack | EW | electronic warfare | M | medium |
|---|--------|-----|-------------------|-----|--------------|
| SOF | special operations forces | TA | target acquisition | mm | millimeters |

**Figure 4-4. Operational system force pool (example)**

4-64. The final combat group structure consists of—

- Command group: reconnaissance and intelligence group. It includes—
  - One reconnaissance battalion.
  - One scout helicopter company.
  - One UAS section (reconnaissance).
- Offensive group: advance group. It includes—
  - One reconnaissance battalion.
  - One SOF company.
- Offensive group: frontline attack group. It includes—
  - Two mechanized CA-BNs.
  - One 122-mm self-propelled gun (SPG) battery.
  - One SPAAG battery.

(See figure 4-5 on page 4-15 for a graphic depiction of a frontline attack group.)

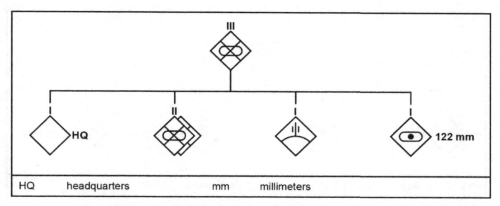

**Figure 4-5. Frontline attack group (example)**

- Offensive group: depth attack group. It includes—
  - Two armored CA-BNs.
  - One mechanized CA-BN.
  - One 122-mm SPG battery.
  - One SPAAG battery.
  - (See figure 4-6 for a graphic depiction of a depth attack group.)

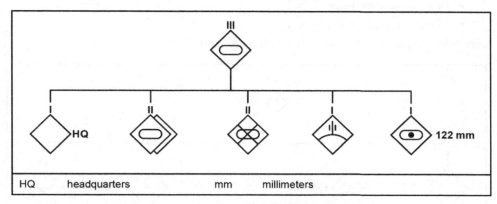

**Figure 4-6. Depth attack group (example)**

- Offensive group: thrust maneuvering group. It includes—
  - Two armored CA-BNs.
  - One 122-mm SPG battery.
  - One SPAAG battery.

(See figure 4-7 on page 4-16 for a graphic depiction of a thrust maneuvering group.)

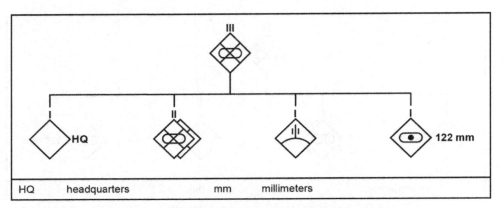

**Figure 4-7. Thrust maneuvering group (example)**

- Offensive group: reserve group. It includes—
  - One mechanized CA-BN.
  - One SPAAG battery.
- Firepower strike group: artillery group. It includes—
  - One 155-mm SPG battery.
  - One heavy MRL battery.
  - One light MRL battery.
  - One UAS section (targeting).

(See figure 4-8 for a graphic depiction of an artillery group.)

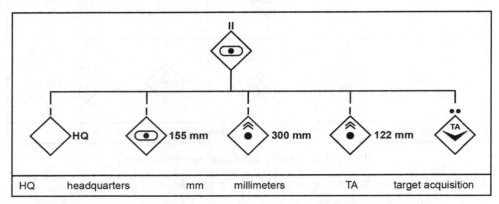

**Figure 4-8. Artillery group (example)**

- Firepower strike group: air group. It includes—
  - One attack helicopter company.
  - One medium lift helicopter company.
- Firepower strike group: mobile artillery group. It includes two 155-mm SPG batteries.
- Firepower strike group: air defense group. It includes—
  - One medium-range surface-to-air missile battery.
  - Two SHORAD batteries.
  - Two SPAAG batteries.
  - One heavy SPAAG or missile battery.
  - One electronic air defense battery.

# Chapter 5

# Tactical Information Operations

This chapter outlines the characteristics and principles that govern People's Liberation Army (PLA) and People's Liberation Army Army (PLAA) information operations at tactical echelons. In keeping with the teachings of Sun Tzu, the PLA considers information operations to be at least as important—if not more important— than maneuver or firepower. Deception, trickery, and concealment are to be employed extensively throughout the information operations campaign in order to manipulate the enemy commander's state of mind, the morale of enemy troops, and the enemy's understanding of the battlefield to the PLAA's advantage.

## OVERVIEW OF PLA INFORMATION OPERATIONS

5-1.   Information operations (IO) are comprised of the wide variety of lethal and nonlethal capabilities with the intent to influence the information environment. This includes information attack and defense, psychological warfare, and deception, trickery, and concealment efforts. The PLA considers IO to be a constant and ongoing effort before, during, and after conflict. Information warfare (IW)—also known as information confrontation—is a subcomponent of IO, and it consists of specific offensive and defensive information actions that are in direct support of military operations. Along with maneuver, firepower, special operations, and psychological warfare, IW is conducted as part of a campaign, usually referred to as campaign information warfare. This is defined by the Chinese as a series of integrated operational battlefield activities that target enemy sensors, information channels, information processing, and decision making. The PLA views IO as a zero-sum activity: they involve both attack and defense, and any information advantages gained in one area may be lost in another. As such, both information attack and defense are considered critical.

5-2.   The objective of PLA IO are to gain information superiority or information dominance. Information superiority requires one to deprive the enemy of information, disrupt the enemy's ability to control or manipulate information of all types, and ensure friendly forces enjoy freedom of maneuver in the information domain. The PLA considers information superiority to be the new high ground of warfare in much the same way as the air domain during the mid-20th century or the land domain prior to that. Moreover, achieving information superiority is thought to have a very low cost-to-effectiveness ratio, achieving significant effects with relatively minimal expenditure in either lives or resources. (See figure 5-1 on page 5-2 for a graphic overview of Chinese IO.)

## CHARACTERISTICS OF INFORMATION OPERATIONS

5-3.   Chinese IO have four primary characteristics: universal permeation, high target value, the importance of integration and synthesis, and the linkage between attack and defense. These characteristics are described in paragraphs 5-4 through 5-7.

### UNIVERSAL PERMEATION

5-4.   IO permeate all domains and are throughout all campaigns. Due to the widespread proliferation of advanced weapons systems, reliance on communications network technology, and information management systems, IO either govern or affect every aspect of modern conflict. Without information superiority, the Chinese believe that they cannot achieve dominance in any other domain, be it land, sea, air, or space. If the information battle is lost, initiative for the entire campaign cannot be achieved, and campaign activities will be fundamentally less efficient or otherwise blunted. IO are an ongoing and continuous process, while IW will likely begin before formal hostilities, making it a precursor to or first shot in active conflict. The conditions to achieve information superiority must be set prior to the commencement of hostilities, must actively support PLA forces during open conflict, and must be maintained through security and stability operations once open conflict has concluded.

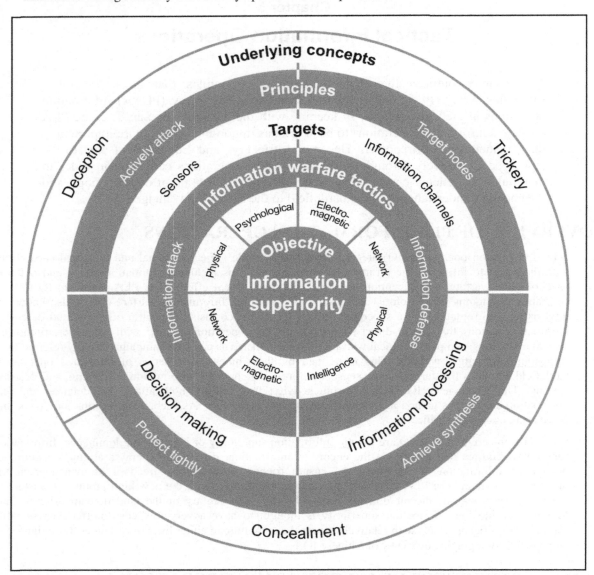

**Figure 5-1. Chinese information operations**

## HIGH TARGET VALUE

5-5.   Information systems are high-value targets. The PLA defines <u>information systems</u> very broadly, including all sensors, information management systems, communications network systems, and decision-making centers at all echelons. System warfare theory suggests that targeting these subcomponents is often an optimal way to undermine the rest of the system, and it is in close keeping with the PLA's warfighting principle of avoiding the enemy's strengths and attacking its weaknesses. Information systems are considered to be the senses and brain of a modern force, and destroying or neutralizing them is an effective way of neutralizing or defeating a stronger enemy. The PLA refers to this approach as paralysis attack or decapitation attack. Attacking enemy information systems is a high-priority mission, while defending friendly information systems is of similar importance. The PLA prescribes a mix of capabilities, both lethal and nonlethal, when targeting enemy information systems.

### IMPORTANCE OF INTEGRATION AND SYNTHESIS

5-6.   IO require a high level of integration and synthesis of a wide variety of systems, personnel, and functions. Effective integration of these varying components is referred to as synthetic quality, and it is one of the unique aspects of IO. Similarly, tactics such as deception, feints, or camouflage must be integrated into the wider information campaign in order to maximize their effectiveness.

### LINKAGE BETWEEN ATTACK AND DEFENSE

5-7.   Information attack and defense must be closely linked. Information attack and defense are viewed as mutually supporting; information attack is most effective when friendly information and information systems are carefully protected, while information defense is enhanced when enemy IO are disrupted by offensive actions. IW is a zero-sum game, and any information superiority gained cannot be undermined by information lost in a different way. Initiative in the information battle is much the same as it is for maneuver or firepower: the side that effectively seizes initiative and then successfully maintains it enjoys a fundamental advantage throughout the conflict. Initiative cannot be either seized or maintained without effective integration of both attack and defense.

## PRINCIPLES OF INFORMATION OPERATIONS

5-8.   Four principles describe the PLA's concept of IO: actively attack, target nodes, achieve synthesis, and protect tightly. These principles are described in paragraphs 5-9 through 5-13.

### ACTIVELY ATTACK

5-9.   Attack is the primary method by which initiative is gained in an IO campaign. The PLA stresses aggressive and integrated attack methods early in the conflict as the best way to ensure information superiority. Further, in an information war it is harder to steal or regain the initiative once it is lost, so greater emphasis is placed on seizing information superiority early in the campaign. Three specific characteristics of IO contribute to the focus on early action: its clandestine nature and ease of concealment; the low requirements on manpower and material resources, making operational sustainment relatively easy; and the inherent vulnerabilities of information systems due to their reliance on the electromagnetic spectrum. A strong initial attack takes advantage of these characteristics, but it requires the conditions to be set long before active hostilities commence.

### TARGET NODES

5-10. The Chinese define a node as a critical component of an information system that either provides a capability or links other nodes. In this context it refers primarily to sensors, information processing centers, and the network backbone that enables them. Nodal attack attempts to identify, isolate, and target these objects, and it is the centerpiece of the IO campaign. Destroying or neutralizing nodes has the highest efficiency of any IW operation: if a critical node is destroyed or neutralized, then all systems reliant on it become either degraded or disabled. Nodal attack embodies the basic principle of targeting points of weakness rather than strength and, if applied appropriately, it can concurrently weaken strong points of the enemy system through isolation and confusion. It is best employed in an integrated manner along with psychological attack in order to maximize confusion and extend the effects of nodal destruction.

### ACHIEVE SYNTHESIS

5-11. Synthesis in the context of an IO campaign implies a variety of different measures all intended to carefully align, synchronize, and coordinate different IW efforts. This includes coordinating information attack and defense to maximize the efficiency of both; assuring deconfliction between different efforts so that one effort does not interfere with another; ensuring that all efforts are unified in their movement towards a clear objective; and focusing on building mutually supportive efforts that in turn create a combined arms effect.

5-12. Adjustment is an analogue to synthesis: it refers to the act of assessing and changing operations quickly and without the need for extensive planning. IO are inherently fluid; changes are rapid and often unpredictable. Agility in operations is vital to ensuring IW efforts are targeting the right objects and achieving desired effects. The primary focus of adjustment is to evaluate different friendly systems and their respective targets, then deconflict actions between them.

## PROTECT TIGHTLY

5-13. As discussed in paragraph 5-9, attack is viewed as the most important basic requirement in campaign IW. The corollary to this axiom is that one must blunt the enemy's attacks in order to effectively gain information superiority. The PLA views information defense as fundamentally more difficult than information attack due to the broad reliance upon—and disparity in—information systems. Because of the wide variety of different points of vulnerability, comprehensive information defense is a practical impossibility—so building systems with resiliency and redundancy is critical. While information defense typically takes the form of passive defense, active defense also has a significant role. This may be described as information counterattack: the use of information attack capabilities to strike at and undermine enemy information attack efforts.

# INFORMATION WARFARE TACTICS

5-14. The PLA breaks IW techniques into two broad categories: information attack and information defense, with each of these containing several subcategories. Information attack is described in paragraphs 5-15 through 5-19. Information defense is described in paragraphs 5-20 through 5-24.

## INFORMATION ATTACK

5-15. The PLA defines information attack as any IW activity intended to weaken or deprive the enemy of control of information. Information attack is the primary means by which information warfare is won, and it is the key to achieving information superiority. There are four subcategories of information attack: electromagnetic attack, network attack, psychological attack, and physical attack. Of note, psychological attack is considered both a form of information attack and its own unique campaign; it is unclear what the relationship is between the two.

### Electromagnetic Attack

5-16. Electromagnetic attack encompasses those activities that manipulate the electromagnetic spectrum to jam, suppress, deceive, or neutralize enemy information systems. There are three subcategories of electromagnetic attack: electronic reconnaissance, electromagnetic suppression, and electromagnetic deception. Electronic reconnaissance attempts to use electromagnetic and other reconnaissance platforms to collect intelligence on enemy information operations platforms, enabling more effective electromagnetic attack operations. Electromagnetic suppression uses electronic warfare capabilities to jam or suppress enemy sensors or communications, softening or neutralizing enemy information systems. Electromagnetic deception manipulates enemy information operations systems—ideally, without enemy knowledge—in support of friendly operations. The PLAA views electromagnetic attack as the centerpiece of most IW operations; this belief is broadly represented in the increasing concentration of IW capabilities at tactical levels, including ground-based and aerial jammers and electronic reconnaissance platforms.

### Network Attack

5-17. Network attack includes those activities that target enemy computer information systems, software, hardware, and their associated networks. Network attack differs from electromagnetic attack in that it is conducted digitally and through networks, rather than through the electromagnetic spectrum. There are two major types of network attack: computer virus attack, which makes use of preprogrammed viruses, and hacker invasion, which uses active code experts working in real time. The aim of both methods is to penetrate enemy information systems. The two techniques can be mutually supportive, and should be closely coordinated. The PLAA broadly prefers clandestine or concealed network attack, wherein the enemy's information is manipulated without its knowledge. This is viewed as the most efficient form of information attack, allowing for more aggressive activities that remain beneath the threshold of active conflict. Concealed network attack operations in peacetime focus on understanding the enemy's operational architecture, disposition, and possible weaknesses, enabling more aggressive network attacks when hostilities commence. Concealed network attacks also focus on acquiring and exfiltrating critical information, particularly that surrounding the defense industrial base. The PLAA seeks to employ network attack even at tactical echelons in order to manipulate enemy situational understanding and trick the enemy into behavior conducive to PLAA tactical operations.

### Physical Attack

5-18. Physical attack comprises those activities that target enemy IO systems with direct physical damage or destruction. At tactical echelons, this primarily entails attacking command and communication centers, network nodes, and sensors. There are three subcategories of physical attack: force strength, which is essentially using maneuver forces to penetrate enemy defenses and destroy the IO systems at close range; firepower assault, which is the use of fires systems of all types—particularly artillery—to target and destroy IO systems at extended ranges; and energy weapons, which involves the use of highly specialized weapons systems to destroy enemy IO systems. The latter includes both current (such as antiradiation missiles) and future (laser or microwave) weapons systems. The PLAA emphasizes firepower assault as the most effective method of physical attack, and firepower assaults against IO systems are a key element of tactical system warfare.

### Psychological Attack

5-19. Psychological attack is composed of those activities that target the enemy's mindset and morale. While categorized as a form of information attack, psychological attack is also a component of psychological warfare. Psychological attack is discussed in greater detail in paragraphs 5-40 and 5-41.

### INFORMATION DEFENSE

5-20. The PLA defines information defense as all those operational activities that protect friendly IW systems from enemy IO. The PLA views IW as a zero-sum activity: any successes gained from information attack are eroded by losses due to inadequate or ineffective information defense. Every potential offensive tactic has a corresponding defensive responsibility. Information attack and defense are viewed as mutually supporting, and they should be carefully integrated.

### Electromagnetic Protection

5-21. Electromagnetic protection contains those measures put into place to resist enemy electromagnetic attack. There are two main modes of electromagnetic protection: counterelectronic reconnaissance and counterelectronic jamming. Counterelectronic reconnaissance is the use of both active and passive means to prevent enemy collection on friendly IW systems. This includes active suppression of enemy collection systems, concealment of friendly electromagnetic signals, the use of decoy or spoof signals to confuse enemy collection, and physical targeting of enemy collection systems. Counterelectronic jamming consists of those systems and techniques that either eliminate or weaken the effects of enemy jamming. This includes hardening IW systems, using of more powerful or more resistant emitters, and carefully monitoring information to weed out disinformation planted by the enemy.

*Note.* The Chinese concepts of electromagnetic attack and electromagnetic defense comprise what both China and the U.S. refer to as *electronic warfare* (EW).

### Network Protection

5-22. Network protection encompasses those measures put into place to resist enemy network attack. There are two main types of network protection: computer virus defense and hacker defense. Each seeks to proactively protect friendly hardware, software, and networks from both overt and covert enemy intrusion. Network defense is both an active and passive activity. Passive defense seeks to prevent, disrupt, or delay intrusion, while active defense seeks to identify and stop intrusion after it occurs. Due to the extensive use of computer networks, along with the wide variety of hardware and software in use, network defense is seen as a highly difficult—though highly important—tactical task.

> *Note.* The Chinese concepts of network attack and network defense comprise what both China and the U.S. refer to as <u>cyber warfare</u>.

### Physical Protection

5-23. <u>Physical protection</u> consists of those activities that defend IW systems from direct lethal attack by enemy forces. There are two categories of physical protection: force strength and firepower, and quality protection. <u>Force strength and firepower</u> incorporates tactical activities that preclude, preempt, neutralize, or defeat enemy direct lethal attack against IW systems. They can be broadly grouped into either defensive and security activities, such as tactical ground defense, or firepower activities, such as counterfire. <u>Quality protection</u> comprises those activities that harden or hide IW systems, making them more difficult to neutralize or destroy. This includes both engineering measures—such as building physical barriers, using underground facilities, or camouflage and concealment—and survivability measures, such as adding armor. The PLAA is very concerned about the vulnerability of IW systems to direct attack—particularly by firepower and energy weapons—and it emphasizes survivability against such attacks.

### Intelligence Protection

5-24. <u>Intelligence protection</u> is IW's counterintelligence component. This includes measures and activities that defeat enemy reconnaissance, surveillance, and espionage, concealing friendly operations and capabilities. Intelligence protection in the form of disinformation can be integrated with information attack—particularly psychological attack—in order to effectively deceive the enemy and trick it into behaving in a preferred way.

## OVERVIEW OF PLA PSYCHOLOGICAL WARFARE

5-25. Chinese psychological warfare encompasses those IO activities wherein a combatant employs information and media in order to target human thought, emotion, and spirit. At tactical echelons, the objective of psychological warfare is to create a psychological condition favorable to friendly forces and unfavorable to the enemy, reducing enemy morale, and consequently, the enemy's will to resist. The PLA views psychological warfare as the operational element of the fundamental reason for conflict: a contest of wills. As such, the PLA notably prioritizes psychological operations. Psychological warfare is considered a fourth operational mode, in addition to land, air, and sea warfare. An effective psychological warfare campaign is considered to be the best possible trade-off, paying a small price in lives and material for as big a victory as possible.

> *Note.* The PLA appears to categorize domains in much the same way as does the United States, but its categorization of psychological warfare as an operational mode is unique. It is unclear how operational modes differ from domains.

5-26. The PLA integrates psychological warfare into all elements of its operations. Psychological warfare is organized as <u>campaign psychological warfare,</u> a discrete campaign in much the same vein as information warfare, maneuver warfare, or firepower warfare. Psychological warfare is considered to have a much broader scope than any other campaign, encompassing not only battlefield combatants, but also national military and economic strength, national will and morale, and national political and social cohesion. Psychological warfare is one of the main reasons why the PLA employs political officers in most of its organizations, and the PLA considers itself to have a fundamental advantage in psychological warfare against nearly any opponent due to its political unity of purpose.

5-27. Psychological warfare seeks to achieve the <u>soft kill</u>: the use of nonlethal coercive means to impose one's will upon an uncooperative opponent. This contrasts with <u>hard-kill</u> techniques, which involve the use of physical force to achieve this coercion. Psychological warfare is thought to have uniquely powerful soft-kill capabilities: a well-constructed psychological campaign is the only thing that can achieve Sun Tzu's definition of supreme excellence—subduing one's opponent without fighting. (See figure 5-2 for a graphic overview of Chinese psychological warfare.)

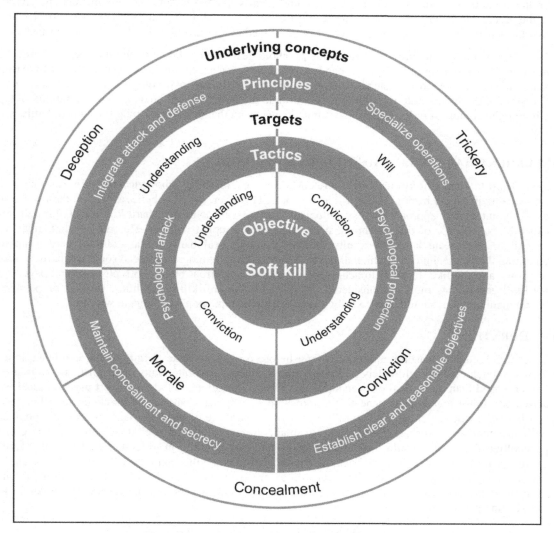

**Figure 5-2. Chinese psychological warfare**

# CHARACTERISTICS OF PSYCHOLOGICAL WARFARE

5-28. Chinese psychological warfare displays four primary characteristics:
- A central role.
- Concurrent hard-kill and soft-kill techniques.
- Long duration.
- Vast scope.

These characteristics are discussed in paragraphs 5-29 through 5-33.

## A CENTRAL ROLE

5-29. Psychological warfare has a central role in modern conflict. The informationized battlefield transitioned psychological warfare from a peripheral or auxiliary military activity to one of the most important operational forms. Due to modern media and social integration, psychological warfare efforts can now target the enemy anywhere and at any time. This includes soldiers on the battlefield, in rear areas, or on the home front; civilians, both enemy and others across the world; and decision makers, including those at the highest level. Psychological warfare must be proportional to the conflict; however, a full-scale deep psychological warfare campaign should not be employed in a localized or limited conflict, for instance.

5-30. Every military activity has some value psychologically and politically, and these second-order effects must be considered carefully by commanders. A campaign's success is no longer measured by casualties, material loss, or territory; instead, it is measured by its overall psychological effect on enemy forces, friendly forces, and neutral or outlying parties around the globe. The ultimate goal of warfare of all types is winning the contest of wills, and this does not always mean that the side winning the tactical battle wins the war.

## CONCURRENT HARD-KILL AND SOFT-KILL TECHNIQUES

5-31. Hard-kill and soft-kill techniques are concurrent in Chinese psychological warfare. Hard kill and soft kill have traditionally been mutually exclusive ideas. However, the PLA believes that modern weapons and a highly integrated approach to psychological warfare can achieve a hard-kill effect through soft-kill capabilities. The use of overwhelming firepower, when coupled with an effective psychological warfare campaign, can create hard-kill results through psychological damage. In addition, new weapons are emerging that are technically nonlethal but have powerful psychological effects: sonic weapons, microwave weapons, and the like. In order to achieve the maximum deterrent effect, both physical and psychological hard-kill capabilities must be considered. More-traditional soft-kill capabilities, such as propaganda and media manipulation, should also be fully integrated into campaign psychological warfare.

## LONG DURATION

5-32. Psychological warfare commences long before formal hostilities begin, and it continues long after they are concluded. It is ultimately tied to a nation's collective morale, unity, and spirit, and so it lasts a far longer period than do tactical or even strategic military operations. Subtle or covert psychological warfare activities—those that fall far beneath the threshold of response—should be practiced long before active conflict commences. These will naturally lean more toward the defense, buoying the friendly population's will and morale, ensuring information advantage in media across the globe, and so on. Campaign psychological warfare should also commence as early as possible in order to set the right information environment; to fool, deceive, and trick the enemy; and to erode the enemy's will. Similarly, psychological warfare operations continue long after hostilities are over. One should attempt to control—or at least influence—the post-conflict historical narrative, while using psychological warfare principles to ensure long-lasting peace and stability.

## VAST SCOPE

5-33. The scope of psychological warfare is vast. Psychological warfare activities stretch across entire regions of the globe, and they can target any individual or organization. The PLA's view of psychological warfare is likely influenced by—or influences—its understanding of Comprehensive National Power: the idea that a nation's strength is not only military and economic, but also political, diplomatic, cultural, and religious in nature. Psychological warfare activities can influence or target anything within any of these fields, and they render the battlefield wholly nonlinear, making coordinated attack and defense not only complicated, but perhaps one and the same. Achieving psychological superiority in conflict establishes a baseline advantage for one's side that permeates to the lowest tactical echelons, and it can enable a technologically or tactically inferior force to prevail in the face of otherwise insurmountable odds.

# PRINCIPLES OF PSYCHOLOGICAL WARFARE

5-34. Four main principles guide psychological warfare:
- Integrate psychological attack and defense.
- Specialize operations.
- Establish clear and reasonable objectives.
- Maintain concealment and secrecy.

These principles are described in paragraphs 5-35 through 5-38.

## INTEGRATE PSYCHOLOGICAL ATTACK AND PROTECTION

5-35. The PLA views campaign psychological warfare through much the same lens as any other combat campaign; in this sense, it includes both offensive and defensive activities. Further, one should target an enemy's weaknesses while attempting to fool or deceive it into attacking one's strengths. Just as with maneuver warfare, success in psychological warfare centers on seizing the initiative, and this is accomplished through attack. Maintaining the initiative is largely a defensive effort. Though different missions, attack and protection are fundamentally linked, and they are mutually supporting. One key difference in campaign psychological warfare is that most psychological protection (or defense) is passive in nature. For example, one cannot hope to prevent an enemy from attempting to influence a population through the global media structure; instead, one must prepare the population to resist these efforts. This process is ongoing and cannot be successfully implemented on a short timeline. Another key idea is that of the psychological counterattack: a targeted effort designed to blunt, disrupt, or marginalize enemy psychological attacks. While broadly similar to other forms of counterattack, the PLA designs psychological counterattacks to target wide audiences and offset enemy psychological attacks on a broad scale, rather than targeting specific enemy operations.

## SPECIALIZE OPERATIONS

5-36. The idea of a specialized psychological operations section is relatively new in both the PLA and the PLAA. In the past, psychological operations generally fell to political officers and nonspecialized troops. Following the model of the U.S. military, the PLAA now builds specialized psychological warfare groups at a variety of different echelons, ensuring integrated support for ground operations

## ESTABLISH CLEAR AND REASONABLE OBJECTIVES

5-37. The wide scope of psychological warfare naturally lends itself to making campaign psychological warfare objectives extremely broad, perhaps excessively so. Planners must carefully assess operational needs and then craft specific, achievable psychological warfare goals. This is particularly true for information and disinformation campaigns. While it is now possible to promulgate a message to a large number of people with relative ease, messages must be at least plausible in order to be of any use. Indeed, messaging viewed as fanciful or clearly false may lead the target to dismiss further messages and sabotage follow-on psychological warfare efforts. The most effective psychological warfare campaigns are those that are focused and that present their targets with believable messages that roughly align with their existing perspectives and biases.

## MAINTAIN CONCEALMENT AND SECRECY

5-38. Deception is ultimately at the heart of nearly every psychological warfare operation. Psychological warfare is one of the only forms of warfare wherein awareness of an enemy's presence decreases the overall effectiveness of the effort. Concealment is at the core of all deception efforts. The source and delivery mechanism for psychological warfare actions should remain opaque to the intended recipient whenever possible. Deception operations are at the core of a number of other PLA and PLAA campaigns, and nearly all of these efforts can be enhanced by effective psychological warfare operations. Secrecy can be considered in much the same vein as concealment: a fundamental component of a successful psychological warfare campaign. Since the main point of psychological warfare is to misinform, confuse, fool, or trick the enemy, any breaches in operational security can conclusively undermine psychological warfare efforts by allowing the enemy to know what is true and what is false.

# PSYCHOLOGICAL WARFARE TACTICS

5-39. The PLA divides campaign psychological warfare into attack and protection, similar to IW. These two aspects result in discrete but mutually supporting operations, and they should be tightly integrated in order to achieve a combined arms effect.

## PSYCHOLOGICAL ATTACK

5-40. Psychological attack contains those activities that target an enemy's situational understanding, conviction, will, and morale. The goal of psychological attack is to control the enemy's awareness and fighting spirit such that its combat power is weakened. The PLAA views this as a supporting effort to information attack, although psychological attack is necessarily broader and less technical in nature. There are two primary methods by which psychological attack is carried out: attacking psychology of understanding and attacking psychology of conviction. Psychology of understanding attacks target the enemy's situational understanding, and they can include everything from camouflage and concealment, to feint attacks or demonstrations, to disinformation efforts. The general objective of these attacks is to lead the enemy into an information trap, a situation wherein enemy leaders and planners make erroneous, faulty, or inefficient decisions due to faulty, insufficient, or misleading information. This can be described as getting inside of the enemy's decision-making cycle, and it is considered one of the most difficult and valuable operational arts in all of warfare. Examples of psychological attack at lower echelons include the use of loudspeakers, simulated smoke or fire, decoys, and unmanned equipment to distort the enemy's situational understanding. Individual soldiers may also be targeted with physical pamphlets or digital messaging in order to upset morale and cohesion.

5-41. Deception is the activity at the heart of psychology of conviction attacks. Practically every PLAA operation places as high a premium on deceiving the enemy as it does maneuver or firepower. Attacks against psychology of conviction target the enemy's fighting spirit, will, and morale. At political echelons, this involves broad campaigns to convince or coerce enemy civilians and politicians to abandon support for the conflict. At tactical levels, the PLAA takes a notably Marxist approach. It seeks to create division between lower ranks and senior ranks by encouraging antiwar sentiments, homesickness, and fear among the enemy's soldiery.

## PSYCHOLOGICAL PROTECTION

5-42. Psychological protection encompasses those activities that attempt to blunt, minimize, or neutralize enemy psychological warfare activities. Psychological protection is divided into two subtypes in much the same way as psychological attack: protecting psychology of understanding and protecting psychology of conviction. Protecting psychology of understanding involves attempting to defeat enemy deception or concealment activities or to minimize enemy disinformation efforts. The objective is to give commanders a clear and accurate picture of the battlefield, enabling quick and accurate decision making, and keeping the enemy out of the friendly decision cycle.

5-43. Protecting psychology of conviction includes those activities that maintain friendly cohesion, morale, and will to fight. It includes defeating broad enemy propaganda efforts or more-direct attacks against friendly soldiers. It also includes political education efforts and other measures designed to enhance esprit de corps, and mental strength and resiliency measures put in place to protect and enhance the individual soldier's psychological well-being. Political officers are also charged with ensuring support and compliance from local civilian populations, thus protecting PLA forces from the impact of political ill-will.

# People's Liberation Army Actions

Part two describes the People's Liberation Army Army's (PLAA's) doctrinal approach to ground combat. The PLAA divides actions into offensive, defensive, and antiterrorism and stability categories, with reconnaissance and security actions performed throughout operations of all types. Chapters six through nine each delve into one of these categories in detail, with discussions on fundamentals, principals, planning, and tactics.

## Chapter 6

# Reconnaissance and Security Actions

The People's Liberation Army Army (PLAA) places a high priority on its reconnaissance and security operations. These are integral to creating and sustaining situational awareness, and they are seen as essential to developing an advantage over one's opponent. System warfare and the application of combat systems are underpinned at all levels by extensive reconnaissance. PLAA maneuver units place a high priority on mounted and dismounted reconnaissance: these units are seen as prestigious, and their soldiers and leaders are highly regarded. Ground reconnaissance remains the chief method by which tactical units gain intelligence on their enemies, but air, space, and electronic reconnaissance are all integrated to varying degrees, even at tactical echelons. PLAA security operations are linked to ground reconnaissance. Their primary intent is to protect main-body formations, give commanders information and decision space, and enable effective risk assessment. The People's Armed Police (PAP) also conduct combat and area security operations, both domestically and in support of PLAA operations. In combat, the PLAA and PAP likely cooperate and coordinate to ensure security throughout the combat area.

## RECONNAISSANCE FUNDAMENTALS

6-1.   Recognizing the changes brought by the informationized battlefield, the PLAA invested substantial time and resources in developing its reconnaissance capabilities. Throughout the Cold War, PLAA units had only light mounted scouts and unreliable manned aircraft to conduct reconnaissance. Despite experiences in the Korean War and the basic tenets of People's War that underpinned the need for effective reconnaissance operations, the PLAA approach during this era was generally to conduct a massive reconnaissance effort using primitive methods and obsolete equipment. The post-reform PLAA, however, places heavy emphasis on advanced reconnaissance methods: more-advanced surveillance platforms, advanced techniques and better training, mechanization and motorization, and intelligence integration. Modern PLAA reconnaissance doctrine is designed to support a sophisticated mechanized force and the system warfare approach.

6-2. PLAA reconnaissance operations obtain information, develop situational understanding, support decision making and actions, and protect key elements of combat power. Reconnaissance and security operations are seen as continuous. Their most important role is providing early and accurate warning of adversary or enemy actions and intent, information about enemy dispositions and mindset, and decision space for commanders at all echelons. The PLAA puts a high priority on reconnaissance operations of all types, considering a commander's situational awareness to be a key factor in being able to detect and exploit enemy weaknesses. Reconnaissance and security missions are typically interwoven with combined arms mission tasks. Reconnaissance is considered an ongoing task that is conducted before, during, and after operations. PLAA leaders match reconnaissance capabilities with missions, seeking to ensure that reconnaissance efforts support coordinated intelligence collection efforts.

6-3. Reconnaissance and security missions are often linked. Reconnaissance can be seen as a proactive task, seeking to develop situational understanding through action, while security is a reactive task, done primarily to preclude enemy actions and enable effective transition to offensive operations. PLAA leaders build reconnaissance intelligence systems to support each mission, task-organizing reconnaissance capabilities that are coordinated with the requirements of the unit's main effort. Dedicated security missions may be assigned to assault groups, combat reserve groups, or defensive groups of all types. These units are designated to screen and protect critical assets from enemy actions. Local security is the responsibility of every unit at all times.

6-4. The PLAA recognizes four key fundamental principles when conducting reconnaissance at all echelons: continuous, command-directed, performed through action, and agile. These principles are described in paragraphs 6-5 through 6-8.

## CONTINUOUS

6-5. Reconnaissance is continuous. Though a reconnaissance mission may be in support of a specific mission or objective, it must commence before and continue after its supported mission. Collection plans and updated reconnaissance objectives ensure an uninterrupted flow of information to the commander, enabling appropriate reactions to enemy activities and changes on the battlefield.

## COMMAND-DIRECTED

6-6. Reconnaissance is command directed. Reconnaissance efforts must be focused on those objectives that support the unit's mission. Reconnaissance forces must be given clear objectives, rules for engagement and disengagement, and follow-on missions. Commanders must ensure that reconnaissance formations are deployed to support established collection efforts.

## PERFORMED THROUGH ACTION

6-7. Reconnaissance is best performed through action. PLAA forces place great emphasis on direct action in reconnaissance—engaging enemy forces in order to develop situations, fix enemy positions, and force enemy commanders into early decisions. Reconnaissance through action also includes the use of airborne sensors to trigger enemy responses and reconnaissance by fire to reveal enemy positions and systems. Direct action also achieves a disruption effect against the enemy, reducing its cohesion and increasing uncertainty.

## AGILE

6-8. Reconnaissance must be agile. The People's Liberation Army (PLA) is universally struggling with a transition from a strong top-down command approach toward a more decentralized methodology. In previous generations, PLAA reconnaissance efforts—particularly patrols—were strictly controlled by commanders, with patrolling elements expected to stick steadfastly to predetermined routes and observation objectives. The reformed PLAA is placing great emphasis on a more flexible approach to patrolling, allowing the patrol leader discretion in supporting the commander's objectives.

*Note.* This publication presents PLA military theory largely as written and prescribed by the PLAA. In most cases this represents a best practice as determined by PLA leadership. Real-world practices of PLA units are largely opaque to outsiders, and they were generally not included as part of the analysis underpinning this document. Moreover, the PLA has not participated in an active conflict in nearly half a century, so real-world applications are minimal. Available information on Chinese military training exercises and the few recent examples of conflict seem to indicate that PLA practices—including those of the PLAA—conform closely to its military theory.

# RECONNAISSANCE AT ECHELONS ABOVE BRIGADE

6-9.   The PLA views reconnaissance as a national-level mission, and it employs national assets in support of reconnaissance operations at all echelons. Examples of national-level reconnaissance assets include satellite surveillance, cyber and electronic surveillance, national information and disinformation campaigns, and national-level special operations forces (SOF) operations. Collectively, these resources typically focus either on strategic-level targets or on deep areas of the battlefield. Tactical commanders may benefit obliquely from surveillance on strategic-level targets, but they are reliant on these assets' contribution to deep reconnaissance.

6-10.   Deep reconnaissance is loosely defined as reconnaissance operations focusing on areas behind that which can be serviced by a given unit's organic weapons systems. For the PLAA combined arms brigade (CA-BDE), this means the areas past the ranges of its own tube and rocket artillery, beginning at somewhere between 35 kilometers (km) and 100 km, and extending through the remainder of the theater of operations. This area typically contains enemy command posts, supply areas, air and sea ports, reinforcement route and staging areas, and long-range fires.

6-11. PLA deep reconnaissance operations consist of a mix of long-range persistent aerial and ground sensors using a variety of manned and unmanned systems, long-endurance SOF and similar light infantry units, and space-based systems. A significant proportion of deep reconnaissance assets are housed in the theater command's reconnaissance and intelligence brigade, while others are national-level assets. Some deep reconnaissance operations are focused on supporting national- and strategic-level assets, while some play an important role in support of tactical operations. Deep reconnaissance capabilities in support of the latter provide targeting for long-range fires, supply commanders with imagery analysis and signals intelligence (SIGINT), and collect on the morale, disposition, and cohesion of large enemy formations. Deep reconnaissance assets are deployed using a combined arms approach, where the strengths of one system offset limitations of another. Collection efforts that cannot be processed in real time, such as imagery analysis, are cross-referenced with collection efforts from SOF and clandestine personnel or information warfare capabilities to create a more holistic intelligence picture than would be possible with a lone system.

6-12. Deep reconnaissance operations feed the PLAA's military planning process during all phases. PLAA commanders may request or be allocated deep reconnaissance assets based on mission requirements. Much of the collection performed in support of tactical echelons is performed by manned and unmanned aircraft, with dismounted ground forces in support. Intelligence is collected in a cycle, wherein collection assets are paired with areas of interest as part of a wider collection plan.

# RECONNAISSANCE AT THE COMBINED ARMS BRIGADE

6-13. Reconnaissance at the CA-BDE is fully integrated with the general intelligence efforts at higher echelons. Intelligence officers are charged with developing an intelligence plan that supports the objective of the commander. Reconnaissance operations are a major component of this plan, especially at tactical echelons. The PLAA employs a simple and robust intelligence cycle designed to integrate all-source intelligence assets with unit operations. PLAA intelligence is intended to be very specific, with discrete questions driving discrete answers.

## COLLECTION PRINCIPLES

6-14. Five principles guide PLAA intelligence collection efforts: mission-supporting, target-oriented, well-planned, well-hidden, and selective. These principles are described in paragraphs 6-15 through 6-19.

### Mission-supporting

6-15. Perhaps more than any other soldier, the reconnaissance soldier must understand the commander's mission and how the reconnaissance effort fits into it. This includes not only understanding the design of the mission, but also its intent.

### Target-oriented

6-16. Intelligence efforts must focus on targets—those objects or activities that best inform commanders and enable their decision-making process. Accuracy is inherent in this principle; inaccurate intelligence is at best useless, or at worst, dangerous.

### Well-planned

6-17. Reconnaissance operations must be well conceived and carefully thought out. This includes identifying objectives, schemes of maneuver, and criteria for engagement, disengagement, and reporting. Conducting reconnaissance without a proper plan results in inconsistent intelligence and compromises the larger mission.

### Well-hidden

6-18. Intelligence operations are at the center of deception, and they are key in gaining an information advantage over the opponent. Carefully guarding intelligence gains, while simultaneously denying or deceiving the enemy's collection efforts, is considered an effort of the highest priority. The intelligence contest is viewed as zero-sum: allowing one's opponent to gain information is detrimental to one's own intelligence operations.

### Selective

6-19. Intelligence efforts must be prioritized; if everything is a focus, then nothing is a focus. Commanders must make clear what their most important pieces of information are, then efforts to match collection assets with targets must mirror this prioritization.

## COLLECTION PROCESS

6-20. The PLAA reconnaissance and intelligence collection cycle follows four steps:
- Develop the reconnaissance plan.
- Deploy reconnaissance forces.
- Screen and integrate the information.
- Report.

The process is considered continuous, beginning again immediately upon completion.

### Develop the Reconnaissance Plan

6-21. This step requires the intelligence officer to identify key terrain features, enemy formations, or other battlefield features that will contribute the most to the commander's decision-making cycle. Intelligence requirements are typically expressed as questions that have clear, unambiguous answers. Good requirements ask and answer questions that are of immediate use to commanders; questions considered of the most value are those which inform the commander about the morale, disposition, and potential weaknesses of the enemy force. PLAA commanders seek to get inside of their opponent's decision cycle, at which point they can disrupt, deceive, or otherwise manipulate the opponent to the commanders' advantage.

## Deploy Reconnaissance Forces

6-22. Reconnaissance forces are deployed in order to answer the questions asked by the reconnaissance plan. Commanders should employ all available reconnaissance capabilities and encourage their reconnaissance forces to employ as many different techniques and technologies as possible. If higher echelon reconnaissance forces are operating in the area, commanders should integrate or cooperate with them. Examples of critical information that reconnaissance forces may seek include enemy strength and disposition, enemy firepower, obstacles, locations of enemy command posts and communication hubs, and key battlefield terrain features.

## Screen and Integrate the Information

6-23. As information begins to flow into the command, the intelligence officer and staff must effectively filter and screen it for accuracy, consistency, and usefulness within the context of the reconnaissance plan. Conflicting reports must be resolved, and incomplete reports may be compared with other information in order to increase accuracy or thoroughness. This phase also emphasizes the importance of security and secrecy in the reconnaissance process: the intelligence officer must ensure that enemy forces do not gain access to recently gathered intelligence, nor be able to effectively collect on PLAA forces.

## Report

6-24. As intelligence is processed, it must be quickly and accurately reported both up and down the chain of command. Reports to higher echelons answer their intelligence requirements, while reports to lower echelons enable prosecution of the commander's intent. Traditionally, the PLAA was very slow to process intelligence, as it was deemed necessary to compute both the tactical and political elements of new information before acting on it. The reformed PLAA places great emphasis on rapid reaction to new data.

# RECONNAISSANCE ORGANIZATIONS AND CAPABILITIES

6-25. The PLAA considers reconnaissance an all-arms occupation, and basic training in reconnaissance, patrolling, reporting, and assessing is provided for all PLAA soldiers. Specialized reconnaissance units, along with SOF, reconnaissance systems, and national-level assets, provide a comprehensive suite of sensors intended to give the intelligence officer a wide-ranging menu of options with which to fulfill intelligence requirements.

6-26. Soldiers and units are the most basic reconnaissance assets. All soldiers are trained on the basics of observation, recording, patrolling, and assessing enemy formations. Collecting information of this type has not historically been a strength of the PLAA. Soldiers were often unable to report developments to their commander due to poor upward communication, a lack of literacy, and poor communications infrastructure. PLAA reforms seek to offset this historic weakness through enhanced training, better-quality recruits, a more decentralized command approach, and an emphasis on bottom-up messaging.

6-27. Observer teams are the most basic dedicated reconnaissance element. They may be trained scout troops or they may be soldiers tasked to perform observation. The observer team functions as a stationary element typically assigned to observe a particular target, most often a piece of key terrain. The observer team then informs its higher echelon command if and when an enemy unit occupies or traverses this terrain. Observer teams are usually not charged with conducting direct action, though they may be given a security mission in addition to their reconnaissance task. Placing and managing observer teams is a basic competency for lower-level PLAA leaders. Observer teams often form the first security line for a deployed unit, screening the unit's main body and giving the unit's commander decision time and space in the event that enemy contact is made. Observer teams may be augmented by unmanned aircraft systems (UAS) or other advanced surveillance systems; simple unmanned aircraft (UA) may be operated by the observer team without additional support.

6-28. The scout team is one of the oldest and most important reconnaissance capabilities in the PLAA. Previous generations relied heavily on the simple two-man dismounted team to provide most ground intelligence—probes to determine enemy disposition, weak points, and avenues of approach were virtually all performed by two-man scout teams. The basic mission of the scout team is the patrol: a movement designed to surveil one or more reconnaissance objectives over a specified geographic area. The scout team remains a key element of reconnaissance; trained scouts are present in all units down to the company level. Scouts today are often mounted, allowing them to conduct patrols over wider areas. They also often have direct communications back to their headquarters element, allowing for real-time updates to commanders. Scout teams may also be augmented by UAS or electronic surveillance capabilities, and many scout teams can operate small UASs without additional support.

6-29. The CA-BDE's reconnaissance platoons, housed within the reconnaissance battalion, provide the bulk of the brigade's long-range reconnaissance patrol capability. The reconnaissance platoon is capable of conducting independent patrols over several days, deploying numerous scout teams over a wide area to perform reconnaissance and security tasks. These platoons operate small organic UA to enhance ground surveillance operations. The reconnaissance platoon may have a direct communications link to the CA-BDE's artillery battalion in order to facilitate rapid artillery fire on time-sensitive targets or to facilitate reconnaissance-by-fire operations. Reconnaissance units also conduct raids, a form of hit-and-run direct action aimed at forcing a confrontation with enemy formations. Raids may be used to reveal enemy positions, to seize enemy prisoners, or to disrupt enemy maneuver. Reconnaissance platoons are likely fully motorized, with heavy CA-BDE reconnaissance platoons operating armored vehicles, such as eight-by-eight light armored vehicles or infantry fighting vehicles (IFVs), in a cavalry-type role. Specialized cavalry fighting vehicles equipped with radars and other sophisticated sensors are being fielded, but there are still of relatively few of them. Reconnaissance platoons work in concert with SOF units, particularly those SOF units assigned to the SOF brigade of the group army. In most cases, the reconnaissance battalion comprises the core of the CA-BDE's reconnaissance intelligence system.

6-30. Each CA-BDE houses numerous UASs, with some in the reconnaissance battalion and some in the artillery battalion. These units operate group two and possibly group three UA, enabling aerial surveillance with a sophisticated suite of sensors in a maximum combat radius of roughly 80 km to 100 km. Sensor options available include simple visual sensors, advanced electro-optical telescopic sensors, infrared sensors, radars, and passive SIGINT sensors. Many PLAA UA are reconfigurable based on mission requirements, allowing commanders to tailor sensor suites to support the collection plan. The reconnaissance battalion's UASs support the ground reconnaissance efforts of the battalion, and they may be task-organized to support reconnaissance platoon operations, particularly long-range or long-endurance patrols. The PLAA also employs UASs in decoy reconnaissance roles: deliberately flying UA in order to entice high-value sensors and air defense shooters to engage and thus reveal their positions. UASs from the CA-BDE's artillery battalion conduct surveillance and targeting operations to support indirect fire. These UASs are likely tied directly to the CA-BDE's artillery systems and provide real-time sensor inputs to enhance artillery targeting. They may also support the CA-BDE's wider collection efforts, but they are more likely employed in support of the brigade's fire support plan, rather than the wider intelligence plan.

6-31. Each group army operates an aviation brigade, which in turn operates a significant number of scout helicopters. Basic aerial reconnaissance has been a key capability of the PLAA for decades, but dedicated scout helicopters are a relatively new addition. It is likely that aviation brigade assets will be task-organized to subordinate CA-BDEs based on mission priority. In many cases, a CA-BDE may be assigned a section of scout helicopters in order to support its collection plans. Scout helicopters are typically light utility helicopters equipped with advanced sensors and limited weaponry, and they can provide direct air-ground communications to CA-BDE leaders as they conduct reconnaissance patrols over wide geographic areas. Scout helicopters may also communicate directly to attack aviation or artillery units, providing elevated forward observation in support of firepower operations.

6-32. The CA-BDE employs limited technical reconnaissance capabilities. These include electronic intelligence (ELINT) and SIGINT systems that support collection efforts through surveillance of the electromagnetic spectrum. PLAA systems of this type were traditionally limited to simple radio direction-finding equipment; ELINT and SIGINT operations were clearly less important than ground patrols. PLAA reforms place much greater emphasis on ELINT and SIGINT, particularly in support of reconnaissance and counterfire operations. PLAA tactical-level ELINT and SIGINT capabilities include radio direction finding, network surveillance, radar detection, and fire finding. These capabilities are rare, and the doctrine and training surrounding their deployment is immature. The systems fielded may vary widely in capability. PLAA commanders also employ decoy systems, designed to spoof, jam, or deceive enemy collection systems, as part of their intelligence operations. CA-BDEs will likely receive significant ELINT and

SIGINT intelligence support from group army collection efforts and regional military capabilities such as dedicated ELINT and SIGINT surveillance aircraft.

6-33. <u>SOF</u> at the group army provide an enhanced long-range ground patrol capability that can be employed either alongside or in addition to the operations of the reconnaissance battalion. Long-range patrolling is considered a key competence for PLA SOF. Small SOF units are capable of conducting days-long patrols both in direct support of CA-BDEs and in deep reconnaissance areas. Many SOF units are equipped with long-range communications capabilities—including satellite communications—enabling them to report back findings in near or actual real time. SOF also train extensively on direct action and reconnaissance by fire. Larger SOF units engage both designated targets and targets of opportunity in order to develop situational understanding through action. SOF units may be tied directly to long-range shooters, including rocket artillery and ballistic missiles, and they can provide precise targeting data for high-value targets in rear areas. These units conduct deep reconnaissance in enemy rear areas. Reconnaissance methods may include passive surveillance, direct action, or covert action.

6-34. <u>Joint and national assets</u> may provide information to CA-BDE commanders and intelligence officers. These assets include rare and highly capable ELINT and SIGINT platforms, satellite surveillance, cyber intelligence, and political intelligence. The degree to which these collection assets are available to the CA-BDE vary significantly. A CA-BDE, for instance, may receive direct support from a space-based asset for a high-priority mission, but it may only passively access national-level intelligence data at other times. National-level assets also contribute heavily to deception campaigns, providing disinformation and decoy efforts that support tactical operations. Wide-area deception operations are considered a very high-priority mission, but it is unclear how effectively they are coordinated with activities at tactical echelons. Other joint assets include manned reconnaissance and strike aircraft operated by air divisions in support of theater commands, national- level SOF, naval surface and subsurface vessels, and clandestine intelligence operations—both domestic and abroad. Of note, the People' Liberation Army Strategic Support Force (PLASSF) is a unified national-level intelligence command, integrating all-source intelligence into a single, easily accessible menu of options presented to lower echelon commanders.

# FORMS OF TACTICAL RECONNAISSANCE

6-35. The PLAA frames tactical reconnaissance in much the same way as the U.S. Army. Ground reconnaissance efforts are tailored to meet specific intelligence collection requirements using one or more forms of reconnaissance. Information superiority at tactical levels begins with a comprehensive and well-executed reconnaissance plan. Tactical reconnaissance efforts are integrated with operational and strategic intelligence efforts to provide the commander with a holistic picture of the battlefield, inform rapid and effective decision making, and deny or deceive enemy collection efforts. The PLAA recognizes two primary forms of tactical reconnaissance: combat reconnaissance and search.

## COMBAT RECONNAISSANCE

6-36. <u>Combat reconnaissance</u> consists of those reconnaissance activities that involve direct action against the enemy. These are implemented before or during a larger operation to inform the commander's decision making as the battle unfolds. Combat reconnaissance is the centerpiece of most PLAA reconnaissance activities, and great emphasis is placed on winning the information battle during aggressive reconnaissance activities. All combat reconnaissance techniques are designed to force the enemy to make a decision—usually related to whether or not to engage—and in doing so, revealing its disposition, strength, or other capabilities. Figure 6-1 on page 6-8 depicts a reconnaissance attack.

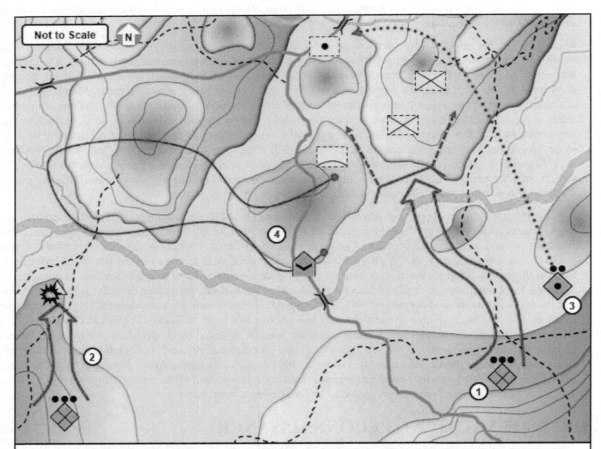

(1) An infantry platoon advances and conducts a reconnaissance attack against an inferior enemy force in order to develop situational understanding and commit enemy reserves. After defeating the enemy unit, the infantry platoon establishes a fighting position. (2) An infantry platoon conducts a raid against an enemy sentry outpost in order to neutralize the outpost and seize prisoners. (3) A 122-mm artillery section conducts reconnaissance by fire against suspected enemy positions, encouraging the enemy to reveal its position and enable friendly counterfire against possible enemy artillery. (4) An expendable unmanned aircraft operates against suspected enemy air defense systems, encouraging the enemy to activate radars or reveal its position via engagement. Artillery is prepared to fire a suppression of enemy air defenses mission as the enemy position is revealed.

**Figure 6-1. Combat reconnaissance (example)**

### Reconnaissance Attack

6-37. A <u>reconnaissance attack</u> is a deliberate or hasty attack on an enemy position performed during a reconnaissance mission. The reconnaissance attack differs from a raid in that it is not hit and run; it is a reconnaissance-in-force that develops into a sustained attack. A commander's orders may grant subordinates the authority to decide, upon encountering an enemy force, whether to develop the situation to gain intelligence and then withdraw, or to press the attack if enemy forces seem weak or otherwise vulnerable.

### Reconnaissance by Fire

6-38. Reconnaissance by fire is the use of fire—typically indirect fire—to either reveal enemy defensive positions or to encourage the enemy to reveal its artillery assets by counterfire. The reconnaissance by fire usually involves delivering a large amount of indirect fire into a specific area believed to be occupied by enemy forces, then observing the area to confirm or deny enemy presence as it reacts to being fired upon.

Reconnaissance by fire may also be used to encourage enemy batteries to react with counterfire, allowing PLAA counterfire capabilities to effectively target them. It has the added effect of depleting enemy ammunition or forcing the enemy to displace due to the counterfire threat.

### Raid

6-39. A raid is a hit-and-run attack, usually performed as a probe to test or locate enemy defenses or dispositions. Raids may also attempt to seize prisoners, either as a primary mission objective or as a target of opportunity. Traditionally, the PLAA was heavily reliant on raids as part of the infiltration tactics it used against strong conventional forces. It is likely that the focus on raids has receded somewhat as more advanced reconnaissance platforms are fielded.

### Feint, Demonstration, Diversion, or Decoy Attack

6-40. Feints, demonstrations, diversions, and decoys employ either expendable assets or careful maneuvers designed to encourage the enemy to reveal dispositions or capabilities through engagement. Feints and demonstrations by ground forces involve intentionally engaging the enemy to fool, deceive, or trick it into making bad decisions. Feints seek enemy contact, while demonstrations seek to avoid contact. Diversions are activities designed to distract the enemy, causing it to focus on one part of the battlefield while secret operations are conducted elsewhere. Decoy operations employ expendable systems designed to fool an enemy into revealing key intelligence. A common example of decoy use is the employment of inexpensive, expendable UA flown over enemy positions, with the intent of sacrificing the vehicles in order to have the enemy's air defense sensors and shooters reveal their positions through engagement. These techniques may also employ cyber or other information warfare elements to manipulate enemy networks, sensors, or mission command nodes.

## SEARCH

6-41. Search consists of those reconnaissance activities that are designed to assess the enemy, terrain, or other battlefield features without deliberate or direct action against enemy forces. The PLAA historically did not emphasize search missions—preferring direct action instead—but on the informationized battlefield, search is recognized as a critical component of overall reconnaissance efforts. The PLAA prescribes that search operations be energetic, active, and well-protected, as reconnaissance forces are high-value targets that are highly susceptible to attack or ambush by a vigilant enemy.

### Linear Search

6-42. A linear search is a form of tactical reconnaissance that focuses on a single route or lengthwise piece of terrain and various features related to that route. Linear search can be thought of as analogous to route reconnaissance in U.S. doctrine. Though any unit can conduct a linear search, for longer routes of sufficient size and condition for CA-BDE use, a reconnaissance platoon is likely required. Intelligence questions that a linear search can answer include condition of roadways, natural or manmade terrain features, presence of enemy forces, and potential hazards along the route. Linear search is considered a key task in support of the advance, the CA-BDE's movement from rear areas to an attack position. The PLAA emphasizes rapid movement and secrecy during approach marches, and high-quality reconnaissance of the route to be used is a major enabler to that end. Linear search may be enhanced by imagery or terrain analysis, map reconnaissance, and aerial surveillance, though scouts moving on the terrain in question are a necessary component to a thorough linear search. Figure 6-2 on page 6-10 depicts a reconnaissance platoon conducting a linear search prior to the movement of a CA-BDE along the specified route.

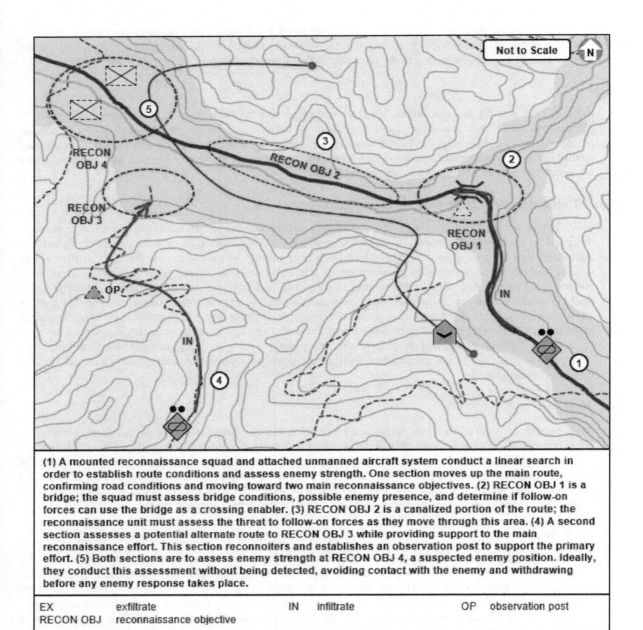

(1) A mounted reconnaissance squad and attached unmanned aircraft system conduct a linear search in order to establish route conditions and assess enemy strength. One section moves up the main route, confirming road conditions and moving toward two main reconnaissance objectives. (2) RECON OBJ 1 is a bridge; the squad must assess bridge conditions, possible enemy presence, and determine if follow-on forces can use the bridge as a crossing enabler. (3) RECON OBJ 2 is a canalized portion of the route; the reconnaissance unit must assess the threat to follow-on forces as they move through this area. (4) A second section assesses a potential alternate route to RECON OBJ 3 while providing support to the main reconnaissance effort. This section reconnoiters and establishes an observation post to support the primary effort. (5) Both sections are to assess enemy strength at RECON OBJ 4, a suspected enemy position. Ideally, they conduct this assessment without being detected, avoiding contact with the enemy and withdrawing before any enemy response takes place.

| EX | exfiltrate | IN | infiltrate | OP | observation post |
| RECON OBJ | reconnaissance objective | | | | |

**Figure 6-2. Linear search (example)**

### Area Search

6-43. <u>Area search</u> is a form of tactical reconnaissance intended to obtain information about a discrete geographic area. This area is typically established by translating intelligence requirements into specific areas on the ground that must be surveilled to confirm or deny intelligence questions. Area search takes longer to plan and execute than other forms of tactical reconnaissance, but it provides detail over a wider area and can answer more questions than other forms. Due to the extended timeline required, area search is best used in support of deliberate operations, where adequate planning and time for execution can be devoted to reconnaissance operations. Area search for the CA-BDE is typically conducted by the reconnaissance battalion, with each reconnaissance troop given a specific area to reconnoiter. Active reconnaissance is conducted by mounted and dismounted scouts supported by UAS, scout helicopters, or electromagnetic sensors; passive reconnaissance is conducted by observer teams. Each subordinate reconnaissance unit is given one or more intelligence questions to answer: examples include confirming or denying enemy presence in a given area; assessing the condition of enemy defensive works, condition, or morale; assessing the condition of key terrain; or finding advantageous routes of attack. Figure 6-3 depicts a reconnaissance troop conducting an area search in support of future combined arms battalion (CA-BN) operations in the area.

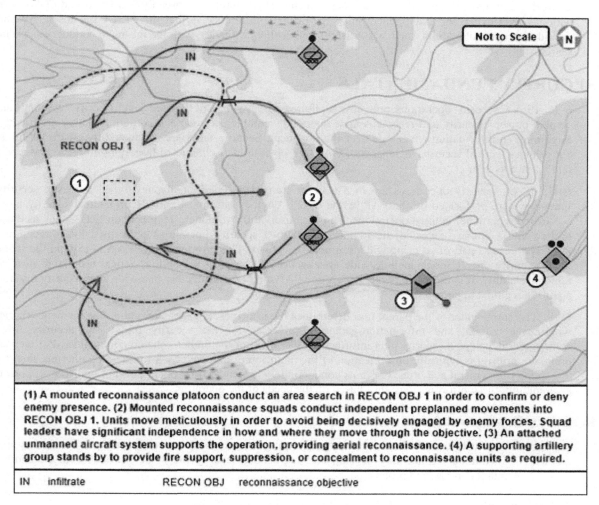

(1) A mounted reconnaissance platoon conduct an area search in RECON OBJ 1 in order to confirm or deny enemy presence. (2) Mounted reconnaissance squads conduct independent preplanned movements into RECON OBJ 1. Units move meticulously in order to avoid being decisively engaged by enemy forces. Squad leaders have significant independence in how and where they move through the objective. (3) An attached unmanned aircraft system supports the operation, providing aerial reconnaissance. (4) A supporting artillery group stands by to provide fire support, suppression, or concealment to reconnaissance units as required.

| IN | infiltrate | RECON OBJ | reconnaissance objective |
|----|-----------|-----------|--------------------------|

**Figure 6-3. Area search (example)**

## Target Search

6-44. Target search is a tactical reconnaissance method by which a reconnaissance element—likely a scout team, forward observer, or scout vehicle—conducts surveillance on a specific target. This can be done to ascertain information about the target or to deliver location data about it to long-range shooters or other engagement elements. Target search is typically passive, as it is intended to feed information to other battlefield systems, rather than to develop the situation through action. As such, it is often conducted from concealment or from standoff ranges. UASs are a key contributor to target reconnaissance, particularly the UAS company that is part of each artillery battalion. These units act in concert with artillery shooters to create a reconnaissance-fires complex, a series of sensor-to-shooter linkages that facilitate rapid target engagement even against over-the-horizon or otherwise concealed target types.

## Electronic Search

6-45. <u>Electronic search</u> or <u>digital search</u> is a relatively new tactic that the PLAA recognizes as a distinct form of reconnaissance, and it has operationalized into long-range or deep electronic reconnaissance. Electronic search broadly refers to the use of collection assets to conduct reconnaissance on enemy signals, networks, and network content, such as social media. This approach is broadly similar to that of Western national-level cyber and intelligence units, the primary difference is that the PLAA intends to try and provide lower echelon units, such as CA-BDEs and artillery brigades, with the means to conduct electronic search.

# SECURITY FUNDAMENTALS

6-46. PLAA security operations protect a supported force with a designated level of early warning and combat power. Security and reconnaissance operations complement one another, and together are the key to gaining an information advantage over the opponent. Security tasks are conducted by all PLAA units, but primary tactical security in support of CA-BDE operations is conducted largely by the reconnaissance intelligence system.

6-47. PLAA security operations are in a state of transition. Traditionally, nearly all domestic security functions were performed by the PAP, while the PLAA focused almost entirely on tactical security tasks associated with major combat operations. Chinese observation of modern regional conflicts—such as those in Iraq and Ukraine—revealed the importance of area security operations performed by conventional military forces as part of securing wider regional objectives. This prompted the PLAA to expand the scope of its security operations to include wide area security in support of stability tactical tasks and, in turn, regional stability. It is unclear how the PLAA and PAP might work together in a wide area security mission against a regional opponent.

6-48. The commander of the parent unit determines the priorities, engagement and disengagement criteria, withdrawal criteria, and information-gathering priorities for the security force. The security force commander develops a security plan informed by the intent of the higher-echelon commander. For tactical units, the priority is typically on protection of the main force, followed by information gathering, then by disruption of opponent activities. For units conducting wide area security operations not in direct support of a parent unit, these priorities may shift based on the needs of the operation.

## SECURITY OPERATION PRINCIPLES

6-49. PLAA security operations revolve heavily around support to maneuver formations. Five principles guide PLAA commanders as they develop and implement security plans: defend stubbornly, standoff, deceive, maintain contact, and protect.

## Defend Stubbornly

6-50. Like most security forces, PLAA security elements will likely face superior opponents as they execute their missions. Commander's guidance to security elements instructs them to fight fiercely even in the face of superior numbers or firepower. This principle is intended to ensure three things: enemy forces are adequately reconnoitered and assessed prior to their reaching their objective, enemy forces are disrupted or demoralized prior to reaching their objective, and friendly commanders are given decision space in the form of time and distance.

## Stand Off from the Opponent

6-51. This principle instructs security commanders to engage their opponents at or near the maximum ranges of their weapons systems. This ensures that enemy forces are engaged throughout their approach to the security line, while making it more difficult to effectively reconnoiter friendly forces. To help enable this principle, security forces are often equipped with longer-range weapons and sensors, such as antitank guided missiles, heavy direct fire guns, and airborne sensors. Effectively standing-off one's opponent is considered a critical step in the counterreconnaissance process, denying the enemy use of key positions or capabilities.

### Deceive the Opponent

6-52. Deception is a common theme in PLAA operations, including security operations. PLAA security commanders are encouraged to deploy their forces in such a way that the opponent overestimates or underestimates PLAA combat power. They are also encouraged to use techniques such as decoys, spoofing, false communications traffic, and feints in order to confuse or mislead enemy commanders. Advanced deception techniques that integrate higher-level capabilities, such as cyber assets, clandestine intelligence personnel, and space-based assets, are being integrated into operations, but their use at tactical levels has not been widely practiced.

### Maintain Contact with the Opponent

6-53. This principle is common throughout army security forces worldwide. Once contact with an opponent is made, security forces attempt to continue it in order to develop the situation through action and gain useful intelligence about the opponent's disposition and intent. Continued contact also ensures that the opponent expends personnel, fuel, and ammunition on the engagement, while further disrupting its plans and timelines. Traditional contact generally revolves around the use of direct fire systems and small arms; advanced techniques may involve the use of information warfare capabilities, deep SOF operatives, or targeting long- range fires by elevated or preplaced sensors.

### Protect the Main Body

6-54. This is ultimately the main objective of security operations, and it should be the task of greatest concern to the security force commander. Security forces conduct stubborn retrogrades in order to buy the main force commander decision space and ensure that the main force does not face a sudden or unexpected attack from an unobserved opponent. Security operations should be integrated with the larger mission of the protected force, enabling the commander to rapidly transition from defensive to offensive operations when the initiative is seized.

## THE SECURITY OPERATION PROCESS

6-55. PLAA security missions broadly follow four steps: assignment, position selection and occupation, security conduction, and retreat. These steps are described in paragraphs 6-36 through 6-59.

### Assign the Mission

6-56. This step is essentially identical to most PLAA operational planning. It involves assessing friendly and enemy forces, the higher echelon mission, coordination measures, and available time. Security missions place a high premium on assessing the enemy's courses of action—specifically the possible directions of attack.

### Choose and Occupy Positions

6-57. Security forces often have some latitude in choosing their positions, and orders may not include specific battle positions for subordinate units. As such, lower echelon commanders may be charged with choosing their own positions based on their best judgment. Positions should be selected based on the expected actions of the enemy, the strength of the position and its ability to blunt or disrupt enemy activity, and how favorable the terrain is to retrograde movement. Positions should be occupied secretly if at all possible, and subordinate units should have a clear concept of when to engage and when to retreat.

### Conduct Security Operations

6-58. There are three primary substeps to any security mission: defend against firepower assaults, defend against direct attacks, and guard against flanking maneuvers. Defending against firepower assaults involves making use of cover and concealment to blunt the enemy's firepower attack, defense against direct attack involves the use of direct and indirect fire to disrupt or destroy enemy offensive actions against the security position, and guarding against flanking maneuvers entails using maneuverable forces to counterattack enemy attempts to target vulnerable flanks of the security position.

**Retreat**

6-59. After meeting their prescribed requirements for delaying the enemy, inflicting casualties, or disrupting enemy attacks, or when their position becomes untenable, security forces should retreat from their security positions. Retreat should be orderly and should include both rear guard actions and concealment actions to protect the main body. If possible, the retreating unit should sabotage any terrain features that may be of use to the enemy. After retreat, the security unit should consolidate and rejoin the combat action alongside the main body. If retreat is not possible, the security unit should conceal itself and then conduct sabotage operations in the enemy's rear areas.

## SECURITY OPERATION TASKS

6-60. The PLAA recognizes four forms of security tasks: combat security, security while marching, security while bivouacking, and garrison security. These tasks are described in paragraphs 6-61 through 6-94.

### Combat Security

6-61. Combat security missions are those activities that actively defend larger units, assets, or key terrain from enemy activity. Missions inherent to combat security include counterreconnaissance and localized defense. Both offensive and defensive missions include a combat security element. There are two primary types of combat security missions: screen and cover.

#### Screen

6-62. The screen is deployed to provide early warning to the main body, to inform the intelligence cycle through action, and to prevent or preclude enemy reconnaissance and intelligence operations. Screening forces may also disrupt and delay the enemy, but they are not intended to become decisively engaged. Counterreconnaissance is the most important mission of a screening unit. The primary targets are enemy reconnaissance, collection, and surveillance assets. Following the reconnaissance and counterreconnaissance operation, screening units typically conduct retrograde operations intended to give the main body commander decision space and inform the commander about enemy intentions. Screens are conducted during both offensive and defensive actions, and they may consist of either dedicated reconnaissance/scout units or nonspecialized units. Screens may operate independently at long ranges from their parent unit, or they may operate in close proximity. Screening units should be capable of conducting independent operations.

6-63. Screens may consist of observation posts; mounted, dismounted, or aerial patrols; or both. The distance from the screen to the main body varies based on unit size, terrain, and the commander's intent. A screen further away from the main body gives greater decision time and a longer period in which to assess the nature of the threat, but it is more difficult to communicate with and sustain. At the CA-BDE, the prescribed distance for a screen is between 5 km and 15 km, though this can vary significantly based on the aforementioned factors. Screening forces typically enjoy a high priority of indirect fire assets, and reconnaissance and surveillance sensors may have direct linkages to shooters to conduct rapid targeting. Indirect fire is considered a critical enabler of retrograde operations, disrupting enemy actions and preventing the screening force from being decisively engaged.

6-64. A screen is typically assigned to the flanks or rear of a main body. Screens can also be placed forward, but this is not employed when the main body is maneuvering; instead, security is provided by forward reconnaissance units as they conduct reconnaissance operations. Screen missions may be employed to close gaps between units, to influence or deceive enemy commanders, or to support wide area security operations. There are two primary types of screen mission: static and mobile. Figure 6-4 depicts a screen operation.

6-65. Static screen missions are most commonly employed during defensive operations. A screening unit is given a geographic area in which to operate, and it is oriented in a specific direction—usually facing away from the main body and toward a potential enemy approach. The screen is given a forward and rear boundary, engagement and disengagement criteria, supporting indirect fire, and reconnaissance objectives in support of intelligence requirements. The screening unit then assesses terrain, forces available, and commander's intent in order to develop a plan to screen the assigned area. The unit integrates multiple sensors, including organic sensors, assets from the force main body, and assets from higher echelon headquarters, to collect and monitor conditions throughout its assigned geographic area. Maneuver forces may also be task-organized with attack or reconnaissance aviation, UA or ground-based sensors, and SIGINT to create an in-depth sensor network. The sensor network facilitates cueing, mixing, and redundancy, providing enhanced sensor coverage throughout the security area. Aerial reconnaissance acts as long-range early warning. Direct and indirect fires and attack aviation are deployed to develop situational understanding through action, to disrupt enemy operations, and to attrit enemy forces. Engineers provide mobility, countermobility, and protection enhancements to the screening unit; obstacles that disrupt or delay an enemy in conjunction with indirect fire are a common approach. The screening unit is responsible for sustainment and support throughout its assigned security zone.

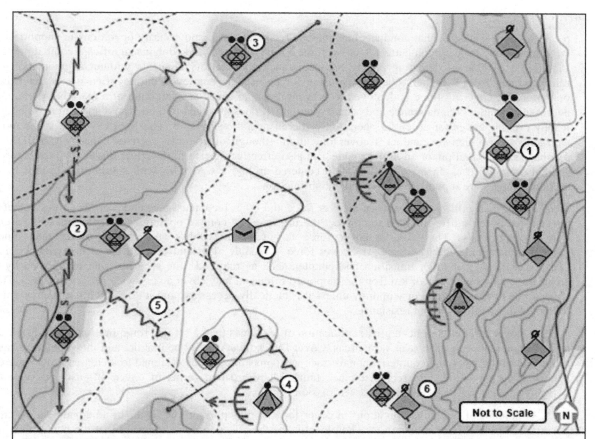

(1) A mechanized infantry company conducts a screen mission 5 km west of the CA-BDE main body. (2) The forward platoon establishes a counterreconnaissance screen and forward fighting position along the front edge of the frontal blocking zone. (3) The rear platoon establishes a depth position, entrenching along the rear of the frontal blocking zone. (4) ATGM vehicles are positioned at key positions in order to stand off the opponent's armor with long-range ATGM ambushes. (5) Obstacles are deployed to slow and canalize enemy vehicle movement. (6) MANPADS teams are deployed throughout the frontal blocking zone to defeat or deter enemy unmanned or attack aircraft. (7) An unmanned aircraft system provides aerial surveillance in support of the entire company.

| ATGM | antitank guided missile | CA-BDE | combined arms brigade |
|------|------------------------|--------|----------------------|
| km | kilometers | MANPADS | man-portable air defense system |
| S | screen | | |

**Figure 6-4. Screen (example)**

6-66. <u>Mobile</u> screen missions are broadly similar to static screens, but they are conducted in support of a main body as it maneuvers. Mobile screens are established using the same control measures as static screens, with the main body commander establishing rough geographic boundaries that support anticipated main-body maneuvers. Mobile screen operations are not typically conducted in front of a maneuvering main body, only to the flanks and rear. The screening force may remain in continuous motion alongside its supported main body, or it may conduct bounding movements with a sister unit.

6-67. Mobile screens are of critical importance in PLAA offensive doctrine. Secrecy in maneuvering is a top priority, a goal that can only be achieved if enemy reconnaissance efforts are effectively neutralized. Mobile screen operations are a key enabler of this goal. Mobile screens may be integrated with screening units from a higher echelon, particularly if a group army is conducting maneuvers over a wide area. Mobile screens may be used to deceive enemy commanders, serving as decoys or distractions to draw reconnaissance efforts and combat power away from the main body.

*Cover*

6-68. A cover is a force that conducts independent reconnaissance and security operations in support of a large main body or geographic area. A cover force is self-sufficient and contains significant combat power. It must be capable of defeating enemy reconnaissance and security elements without augmentation or reinforcement from a higher echelon. Covering forces may be employed during offensive or defensive operations. A covering force is deployed to prevent reconnaissance and targeting of a main body through aggressive counterreconnaissance and security actions. It does not need to be tied to a specific main body; it may provide cover for an entire operation or be an operation in itself. The light CA-BDE is the most well-suited formation to conduct a cover operation—though other units, such as a heavy CA-BDE, a reconnaissance battalion, or an SOF brigade, can also effectively perform the cover mission. A cover force is likely organized as an operational system centered around a CA-BDE, with subordinate groups built around defense, fire support, antitank, and antiair missions.

6-69. The primary difference between a cover force and other reconnaissance and security forces is that a cover force is self-sufficient, able to conduct the full range of combat operations independent of a parent unit or main body. This gives a cover force far greater flexibility and combat power than other reconnaissance and security forces. The cover force is capable of extended defensive or delaying actions, and it can conduct its own transition and counterattack as required. The security zone established by a cover force may be dozens of km from its supported force and may cover a very wide geographic area. The cover force does not retreat or withdraw unless it is tactically necessary, and it does not allow enemy units to bypass its position without resistance.

6-70. A cover force represents a greater allocation of resources than a screen. It requires extended logistics and reconnaissance support over wide areas. Cover forces must usually be mobile, and they contain a high proportion of antitank and counterreconnaissance weapons and systems designed to increase combat power against enemy scouting and reconnaissance units. There are two forms of cover: offensive cover and defensive cover. Figure 6-5 depicts a cover mission.

6-71. In support of offensive operations, a cover force may deploy on the flanks of an advance or to any potentially vulnerable area of terrain. The objective of the cover force is to ensure that the main effort of the attack does not suffer from enemy surveillance, direct attack, or indirect fire from the force's assigned security area. This is accomplished through the deployment of the force to impede enemy movement, defeat or degrade enemy reconnaissance capabilities, and defeat enemy units that enter the security area. A cover force may also deploy to the front of an advance, operating at an extended distance from the main body. When operating in this capacity, the cover force aggressively seeks weaknesses in the enemy position and exploits them as they appear, enabling follow-on operations into enemy rear areas.

6-72. Offensive cover units may have to rapidly displace to new security areas in order to properly cover a maneuvering main body. If operating at a significant distance from the supported force, integrating maneuver and reconnaissance operations with the main body is a challenge. Cover forces should generally be considered independent and be allowed to conduct operations within prescribed geographic limits without excessive input from higher echelons.

6-73. In support of defensive operations, a cover force protects a specific flank or geographic area from enemy attack and surveillance. The cover force's security zone extends across the full width of the vulnerable area, and the cover force is responsible for reconnaissance, counterreconnaissance, and security throughout. The cover force compels the enemy to reveal its disposition and intent, enabling an effective counterattack and transition to offensive operations. It first looks to transition to offensive operations by itself. If this is deemed impossible, then a counterattack force from the main body may be committed.

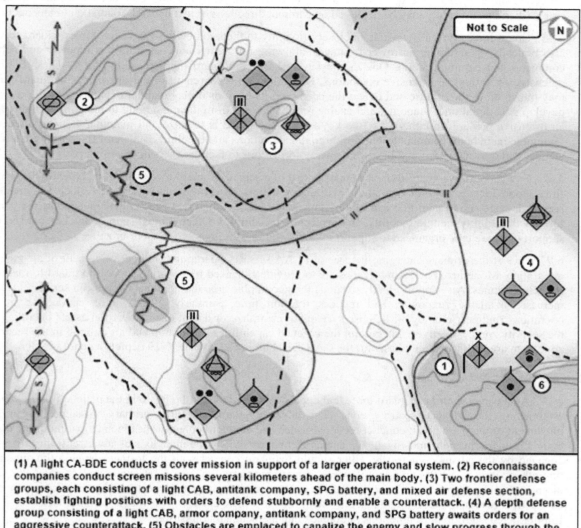

(1) A light CA-BDE conducts a cover mission in support of a larger operational system. (2) Reconnaissance companies conduct screen missions several kilometers ahead of the main body. (3) Two frontier defense groups, each consisting of a light CAB, antitank company, SPG battery, and mixed air defense section, establish fighting positions with orders to defend stubbornly and enable a counterattack. (4) A depth defense group consisting of a light CAB, armor company, antitank company, and SPG battery awaits orders for an aggressive counterattack. (5) Obstacles are emplaced to canalize the enemy and slow progress through the defensive zone. (6) A firepower group consisting of a rocket battery and heavy gun battery provides fire support throughout the combat area.

| CAB | combined arms battalion | CA-BDE | combined arms brigade | S | screen |
| SPG | self-propelled gun | | | | |

Figure 6-5. Cover (example)

## Security While Marching

6-74. Security while marching encompasses those missions that protect a main body while it is on the move. These missions are collectively referred to as guard missions. Guard is a security mission to protect the main body by engaging and standing off the enemy, giving the main body decision space and reaction time. Guard missions differ from screens in that guard units are expected to actively protect the main force from a significant enemy attack, rather than simply delay or disrupt it. The primary mission of a guard force is to stand off enemy forces in order to prevent observation, direct fire, and indirect fire observation on the main body. The guard force also conducts reconnaissance and counterreconnaissance operations within its combat area, seeking to destroy or neutralize enemy reconnaissance assets while informing friendly reconnaissance objectives. The guard force must be prepared to conduct a decisive engagement, either to defend the main body from a superior force, or to counterattack and defeat a weaker opponent. There are three types of guard missions: advance, flank, and rear. In a noncontiguous situation, advance, rear, or flank guard forces may be deployed on cardinal directions or in a general orientation to an enemy.

6-75. Guard forces typically operate in closer proximity to the main body than screen forces. Guard forces also typically operate over narrower frontages, allowing for greater concentration of firepower and better command and communication. One of the key tasks of the guard mission is transition, the shift from defensive operations to offensive operations. Depending on the strength of the opponent, the guard force may transition to the offense and defeat the enemy on its own, or it may maintain contact with the enemy, develop situational understanding, and enable reinforcements from the main body to transition rapidly into an offensive posture. In either case, the key task of the guard is to seize and maintain tactical initiative, giving the main body commander the ability to dictate actions and timeline. The guard force may conduct a tactical withdrawal or lead, fix, or disrupt enemy formations to facilitate seizing the initiative.

6-76. Fire support to a guard force is similar to that of a screen. A guard force, however, typically has more fire support for a given unit size, and it will also integrate observation and aerial surveillance to create a reconnaissance-fire system that can rapidly engage targets with indirect fire. Engineers support the guard force with mobility, countermobility, and protection capabilities. Other combat support and combat service support units are task-organized to support guard operations throughout the security zone.

6-77. One of the primary emerging focuses of PLAA security operations is the development of air-ground integration. Modernized aviation brigades are to provide enhanced rotary-wing support to guard formations in security zones. Air support greatly expands the geographic area that the guard force can service, greatly increases available firepower, and reduces reaction time. Similarly, PLAA units conducting guard operations routinely employ air defense systems to neutralize or defeat enemy aviation assets operating in the security zone. Effectively integrating the twinned capabilities of aviation and air defense in the security zone is an area of tactical emphasis for the CA-BDE. Figure 6-6 on page 6-19 depicts a guard mission.

### Advance Guard

6-78. An advance guard is a guard force that deploys ahead of a main body along the main axis of advance, establishing the advance security zone. The advance guard has three primary missions: streamline movement of the main body; conduct reconnaissance and counterreconnaissance in anticipation of the main body's primary mission; and engage the enemy if it appears along the main axis of advance. If the enemy is engaged by the advance guard, the latter should maintain contact with the former, develop the situation through action, and transition to offensive operations based on enemy strength and engagement and disengagement criteria. The advance guard likely represents the largest proportion of combat power for the CA-BDE's reconnaissance intelligence system, and it works closely with the frontline attack group or frontier defense group during periods of contact, engagement, and transition.

6-79. When functioning as part of an offensive group, the advance guard should maintain close contact with reconnaissance forces operating to the front and flanks of the main body. These units inform the advance guard about enemy dispositions and intent, enabling the advance guard commander to make rapid and effective decisions about positioning and the allocation of fires and reconnaissance assets. The advance guard commander remains in close contact with the main body commander in order to facilitate rapid decision making and transition to the offense. Typical advance guard force tasks while on the offense include deceiving the enemy about the main body's intent and disposition; forcing the enemy to commit reserves early; and targeting critical assets, such as command and communication nodes, fires systems, and lines of communications.

6-80. When functioning as part of a defensive group, the advance guard should deploy along the enemy's most likely avenue of approach. The primary mission of the advance guard in this situation is to create a disruption zone, with the objectives of slowing the opponent's progress, developing intelligence through reconnaissance by force, deceiving the enemy as to the main body's disposition and intent, and forcing the enemy to commit to one course of action. Once the guard force is in contact, it defends, delays, and disrupts in support of the main body force. Tactical offensive actions, such as ambushes, raids, or limited counterattacks, may be employed while part of a defensive group in order to facilitate the transition to offensive operations.

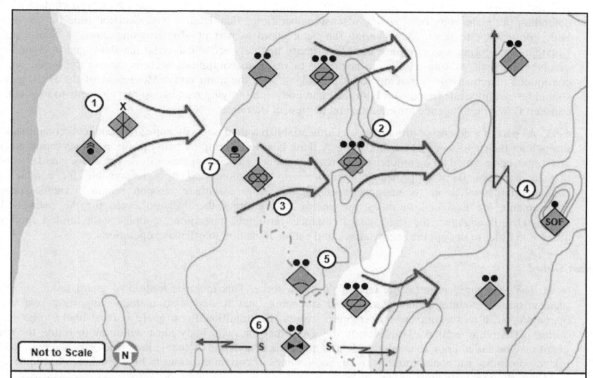

(1) A light CA-BDE conducts an advance to an assault position. The reconnaissance intelligence group is given the task of security while marching. It employs an advance guard, flank guards, and rear guard. This graphic depicts the advance guard, which is built around a mechanized reconnaissance company, mechanized infantry company, 122-mm SPG battery, attack aviation section, and MANPADS squad. In addition, the group army provides dismounted scout patrols and SOF patrols in deep areas. (2) The three platoons of the reconnaissance company each advance on their own axis, with orders to scout the area and report any enemy presence without being decisively engaged. (3) The mechanized infantry company forms the advance guard's reserve and assists any reconnaissance platoons in breaking contact. (4) Group army dismounted patrols and SOF conduct deep reconnaissance in front of the advance. (5) MANPADS sections stand off enemy attack aviation and target any enemy unmanned aircraft that surveil the advance. (6) The attack helicopter section screens the southern flank, hunting unmanned aircraft and providing reconnaissance to the south. (7) The advance guard's indirect firepower consists of a 122-mm SPG battery, augmented by the CA-BDE's rocket battery.

| CA-BDE | combined arms brigade | MANPADS | man-portable air defense system |
|--------|----------------------|---------|--------------------------------|
| S | screen | SOF | special operations forces |
| SPG | self-propelled gun | | |

Figure 6-6. Guard (example)

### Flank Guard

6-81. The flank guard protects the flank of the main body, establishing the flank security zone. Flanks are broadly defined as the areas parallel to the axis of the main body's direction of travel or, while stationary, areas not along the expected enemy main avenue of approach. A flank guard will generally have more geographic territory to cover than an advance guard, and it will do so with fewer forces. Thus, flank guards must have enhanced wide area reconnaissance capabilities and mobility, but they will likely have less overall combat power as compared to advance guards. The flank guard should stand off at sufficient distance to give the main body commander adequate decision space in the event of a flank threat, but it must maintain constant contact with the main body at all times.

6-82. Flank guards that are part of offensive groups move along with the main body, maintaining continuous surveillance of their assigned geographic areas while simultaneously conducting counterreconnaissance and security operations. If contact is made, the flank guard maintains contact while defending the main body from enemy offensive operations; this defense buys decision time for the main body commander to react. Maneuvering the flank guard as part of offensive operations is challenging. Bounding or continuous march are the two primary methods of maneuvering the flank guard. Bounding involves one unit passing another in sequence to maintain contiguous security across the flank, while continuous march involves moving constantly along with the main body. Movement of the flank guard should be carefully integrated with both the main body and outlying reconnaissance elements to maintain a constant flow of intelligence information and to prevent fratricide.

6-83. As part of a defensive group, flank guards establish a flank security zone and conduct reconnaissance throughout their assigned geographic areas. A flank guard may employ battle positions, observation posts, and ambushes enabled by a comprehensive reconnaissance and surveillance effort. The flank guard ensures that its assigned flank is protected against enemy intrusion, and it is responsible for conducting counterreconnaissance in its assigned area. Deployments maximize weapon ranges. Extended-range engagements are designed to delay the enemy and prolong the decision cycle for the main body commander. If engaged, the flank guard conducts retrograde operations, coupled with limited spoiling attacks, in order to disrupt the enemy attack and enable transition to offensive operations.

### Rear Guard

6-84. The rear guard protects the rear of the main force. This includes traditional guard tasks, such as reconnaissance, counterreconnaissance, and screening, and it may also include supporting rear area operations, such as sustainment and engineer efforts. The traditional rear guard is most likely to be used during offensive operations. However, in the event that the main body must withdraw or retire, the rear guard may be called upon to conduct a delaying or screening action to enable the maneuver. The rear guard has responsibility for maintaining the main body's lines of communications to higher echelon headquarters and supporting security units in rear areas.

## Security While Bivouacking

6-85. Security while bivouacking, also called security at the halt, includes those security activities that protect a main body when it has stopped moving, but it is not located in a garrison or rear area. These missions are designed to protect the main body from enemy reconnaissance and attack. It is up to commanders to balance the security needs of the unit with available forces by carefully assessing the nature of the threat, the vulnerability of the main body, and the number of troops available to conduct security missions. There are two types of security missions conducted while bivouacking: patrol and sentry. Figure 6-7 on page 6-21 shows security while bivouacking.

### Patrol

6-86. Patrol missions, discussed in paragraph 6-28, are movements designed to facilitate surveillance of a given geographic area through the use of small, mobile scouting teams. For a main body at the halt, patrols should conduct wide-ranging assessments of the surrounding terrain, then maintain a reconnaissance presence ahead of the security zone. These patrols should locate and fix potential enemy activity targeting the main body, allowing the commander decision space to commit security or defensive group forces. In general, patrols are not to take direct action against enemy forces and should instead rely on cover, concealment, and deception to break contact. However, if a patrol encounters an enemy patrol of similar size and strength, the patrol should either destroy or force the withdrawal of the enemy unit in support of the wider counterreconnaissance mission.

### Sentry

6-87. Sentry missions are security missions that employ small, static units carefully positioned in and around the security zone in order to provide advanced warning of enemy activity; reconnaissance and counterreconnaissance support; and, if necessary, early engagement of enemy forces. Sentry missions are similar to patrols, except that they remain in place or move comparatively little. Sentry missions are conducted by observer teams, which should be carefully concealed while occupying positions of tactical advantage. Observer teams should have direct communications links to their supported headquarters.

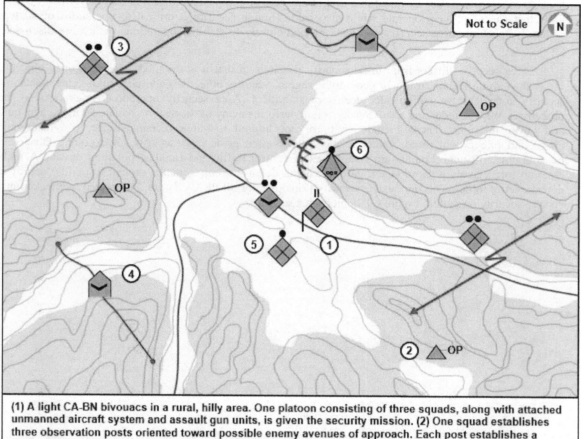

(1) A light CA-BN bivouacs in a rural, hilly area. One platoon consisting of three squads, along with attached unmanned aircraft system and assault gun units, is given the security mission. (2) One squad establishes three observation posts oriented toward possible enemy avenues of approach. Each post establishes a direct communication link back to the battalion headquarters, allowing it to rapidly report on any intelligence developments. (3) One squad, split into two sections, conducts patrols along the edge of the security area. The sections stand off enemy reconnaissance elements, screen the main body, and surveil their respective patrol areas. (4) An unmanned aircraft system section with two aircraft operates in support of the patrolling squads. (5) One squad is held as a reserve group. This units reacts to developments in the security area and reinforces sentries or patrols under attack. (6) An assault gun is positioned in an ambush location, anticipating the enemy's most likely avenue of approach.

| CA-BN | combined arms battalion | OP | observation post |

**Figure 6-7. Security while bivouacking (example)**

## Garrison Security

6-88. <u>Garrison security</u> operations are those security operations that take place behind frontline units. These missions differ from security while bivouacking missions in that they take place in more permanent areas: in and around supply depots, assembly areas, air and sea transport facilities, and in civilian areas. This security mission should not contact enemy frontline units, and so it focuses heavily on defeating small-scale enemy penetrations, civilian disruption in rear areas, and air intrusions. The PLAA recognizes the importance of wide area security as part of its transition to an informationized force, and it is developing capabilities that enhance this mission. In many cases, the PAP will play an important role in garrison security operations, freeing up PLAA resources for more-demanding missions.

*Area Security*

6-89. <u>Area security</u> is conducted to protect friendly forces and critical assets across a specified geographic area. Area security can be assigned to a unit of any size or composition, but it is best performed by mobile units that can effectively operate while dispersed. Though security is critical to operations of all types, generally speaking, area security missions are tied closely to stability operations, which are discussed in chapter 9.

6-90. Area security differs from other security missions in that it is oriented around <u>security objectives</u>, rather than around a unit, route, or geographic region. Security objectives are those high-value assets in the security zone that the commander deems most critical. Area security operations orient around these objectives. When conducting area security, a unit performs many of the tasks associated with screen, guard, and cover missions. These include—but are not limited to—area reconnaissance, fires integration, air-ground integration, route reconnaissance, observation posts, and security checkpoints. Observation posts and security checkpoints are established to control traffic and canalize potential threats. A mobile reaction force, likely a depth attack group or depth defense group, remains on call to rapidly deploy and defeat enemy forces. The same basic security principles for all PLAA forces are applied to checkpoints in an area security operation. Figure 6-8 depicts area security.

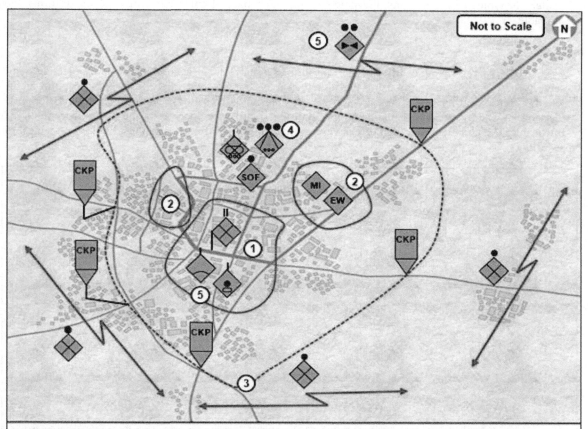

(1) A light CA-BN conducts an area security mission in a low-density urban area. The CA-BN is organized into a frontier defense group, a depth defense group, and a firepower group. (2) The CA-BN commander establishes two security objectives: a hospital complex and an annex housing an electronic warfare team and a military intelligence team. (3) The frontier defense group consists of three light infantry companies, which maintain several checkpoints around the security area and conduct regular squad patrols around the perimeter in order to provide a strong visible presence. (4) The depth defense group consists of an attached mechanized infantry company and assault gun platoon. This group reinforces any part of the security perimeter that comes under attack. A SOF team provides reconnaissance and sniper support throughout the security area. (5) The firepower group consists of a 122-mm SPG battery, a SHORAD battery, and an attack aviation section. This group provides artillery and air defense firepower, along with an aerial reconnaissance and attack capability.

| CA-BN | combined arms battalion | CKP | checkpoint | EW | electronic warfare |
|-------|------------------------|-----|-----------|-----|-------------------|
| MI | military intelligence | SHORAD | short-range air defense | SOF | special operations forces |
| SPG | self-propelled gun | | | | |

**Figure 6-8. Area security (example)**

6-91. Area security nearly always involves a <u>perimeter</u>, a boundary around the security zone that represents the limit of the area security operation. The unit assigned to the area security mission is responsible for creating and maintaining the perimeter. The perimeter may focus either on keeping enemy forces out of the security area or on keeping persons of interest within, for example, a cordon around a separate operation. Unlike other security operations, area security may require a strong visible presence. Instead of seeking to conceal friendly forces and deceive an enemy, a unit conducting area security may choose instead to prominently display its combat power to dissuade or deter enemy actions within the perimeter.

6-92. Area security is likely to be a peacetime or transition task, and it is actively conducted within Chinese borders. This requires integration with PAP and local police forces, as both of these organizations may have roles in maintaining security in a given geographic area. PLAA involvement in domestic security matters is typically limited, but the potential for an expanded role following ongoing reforms seems probable.

### Route Security

6-93. Route security is a security mission performed along a specific route. The main objective of route security is to ensure that units using the route can move rapidly and safely from point to point along its length. Route security operations are terrain oriented and defensive in nature. They typically involve clearing of enemy presence, maintaining a constant patrol presence, ensuring that the route remains trafficable, and conducting limited offensive operations to stand off enemy actions. Figure 6-9 depicts route security.

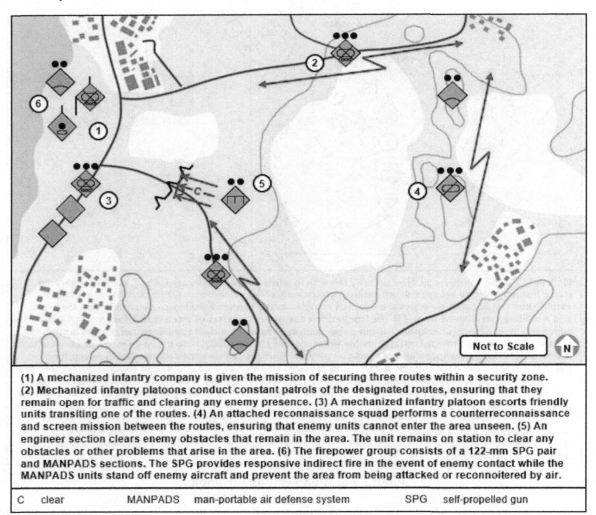

(1) A mechanized infantry company is given the mission of securing three routes within a security zone. (2) Mechanized infantry platoons conduct constant patrols of the designated routes, ensuring that they remain open for traffic and clearing any enemy presence. (3) A mechanized infantry platoon escorts friendly units transiting one of the routes. (4) An attached reconnaissance squad performs a counterreconnaissance and screen mission between the routes, ensuring that enemy units cannot enter the area unseen. (5) An engineer section clears enemy obstacles that remain in the area. The unit remains on station to clear any obstacles or other problems that arise in the area. (6) The firepower group consists of a 122-mm SPG pair and MANPADS sections. The SPG provides responsive indirect fire in the event of enemy contact while the MANPADS units stand off enemy aircraft and prevent the area from being attacked or reconnoitered by air.

| C | clear | MANPADS | man-portable air defense system | SPG | self-propelled gun |

**Figure 6-9. Route security (example)**

6-94. Route security also typically involves escort, particularly of vulnerable targets such as convoys or command vehicles. Units assigned a route security mission may conduct escort missions as part of route security, or they may enable other units to conduct them. Escort missions are performed in much the same way as any other tactical movement. The route is reconnoitered, cleared of potential obstacles, then screen or guard forces stand-off enemy units at sufficient distance that the latter cannot target the assets moving along the route.

# Chapter 7

# Offensive Actions

Chinese forces consider tactical offensive actions to be the decisive form of land operations. The active defense strategy relies on effective and credible tactical offensive actions as the basic contribution of ground forces. Tactical offensive actions destroy an opponent's will to fight through a combination of firepower, maneuver, deception, and information warfare. This chapter outlines the People's Liberation Army Army's (PLAA's) methodology for planning, preparing, and executing tactical offensive actions.

## OVERVIEW OF PLAA OFFENSIVE OPERATIONS

7-1.   Throughout its history, the People's Liberation Army (PLA) has emphasized offensive operations. The idea that war can only be won by attacking is fundamental to People's War theory, and it has been tested time and again throughout the Chinese Civil War and the Korean War. PLA leaders historically emphasized the spirit of the attack—what Westerners would call *élan* or *esprit de corps*—as the only way their military forces could overcome the technological, firepower, and training superiority of their opponents. Developing this spirit in their formations was a fundamental skill of PLA leaders. Indeed, the role of political officers and commissars was in large part to help develop a unity of purpose underpinning the spirit of the offensive.

7-2.   The PLAA today takes a similar view toward offensive operations, though with a greater focus on firepower, joint integration, and maneuver instead of the more traditional approach of infiltration followed by close combat. In keeping with Mao's principles, PLAA operations focus on destroying enemy formations rather than taking ground. As such, objectives are often described as enemy formations. PLAA forces seek to use a mix of maneuver, deception, and firepower to preclude enemy actions, isolate enemy units, and then fight the isolated enemy to annihilation. PLAA units integrate advanced deception and information warfare capabilities to fix enemy forces and then conduct decisive attacks on enemy weak points. They employ firepower not only as an enabler of maneuver, but also as an offensive tactic in itself: employing massed fires to destroy, neutralize, or fix opponents.

7-3.   PLAA offensive operations are performed to accomplish one or more of the following objectives:
- Destroy, defeat, or neutralize enemy formations, personnel, or equipment.
- Enable friendly freedom of maneuver.
- Restrict enemy freedom of maneuver.
- Gain information.
- Gain control of key terrain.
- Disrupt enemy operations.

PLAA leaders traditionally preferred to operate using a grand battle plan—an extensively planned and prepared operation that demanded adherence to a rigid hierarchy and a complex, often inflexible plan. This approach was well-suited for armies consisting largely of undertrained and underequipped conscripts, and leaders who emphasized party loyalty and political enthusiasm over tactical competence. PLA reforms are attempting to significantly change this approach, employing a decentralized approach to leadership, greater tactical flexibility in planning, and more empowerment and better resourcing of leaders at lower echelons.

# THE INFORMATIONIZED BATTLEFIELD AND OFFENSIVE OPERATIONS

7-4. The PLAA recognizes six important trends that significantly impact offensive operations on the informationized battlefield: transparency, precision munitions, electronic warfare (EW), operational tempo, multidimensional battle, and cost. Leaders and planners must account for these trends when conducting all types of offensive operation.

# THE TRANSPARENCY OF THE INFORMATIONIZED BATTLEFIELD

7-5. Advanced multispectrum surveillance, networked intelligence, and the connected world have combined in such a way that true deception is much harder to achieve than it was in the past. Units will find it far more difficult to assemble, move, and prepare for offensive operations without the enemy uncovering their intent. Commanders are instructed to integrate deception plans across the entirety of the information spectrum, and they must always assume that they are being surveilled by a clever enemy.

## THE PROLIFERATION OF PRECISION MUNITIONS

7-6. Precision munitions enable rapid, deep, and precise targeting of critical assets all across the informationized battlefield. Firepower no longer requires mass; effective synchronization of precision munitions can achieve the same effect that formerly required dozens of attack sorties or hundreds of guns. Precision platforms usually overlap one another to achieve a combined arms effect, and they are enabled by advanced multispectrum surveillance in support of targeting. Commanders must always assume the enemy can strike them. Commanders must work to reduce vulnerability by protecting forces both physically and electronically; by neutralizing or defeating attacks by precision munitions; by moving rapidly; and by deceiving the enemy to throw off its targeting process.

## THE IMPORTANCE OF ELECTRONIC WARFARE

7-7. The PLAA not only puts a high priority on its own EW capabilities; it also anticipates enemy capabilities will contest its network and communications capabilities. EW has the potential to offset precision munitions, sensors, and joint communications networks, creating an environment where aggression and short-range firepower can prove decisive. As this is equally true for both Chinese forces and their opponents, electromagnetic protection is a high-priority mission for the PLAA. Chinese leaders and units are instructed to train in communications blackout conditions, relying on ingenuity and tactical competency to overcome the effects of communications isolation, and wherever possible, use communications means that are not susceptible to enemy EW efforts.

## THE RAPID TEMPO OF OPERATIONS

7-8. The informationized battlefield moves quickly, and changes in the environment can be unpredictable and sudden. Motorized and mechanized ground units, along with aerial units, can traverse significant distances in short periods of time. Firepower systems can range distant targets, and can they react quickly to targeting information. All of these developments likely mean that windows of opportunity for commanders to seize the initiative are getting smaller and shorter. Commanders must be mentally agile to take advantage of these windows in the absence of direction from higher echelons, particularly if the situation deviates from planned one. Comprehensive planning and training helps to ensure that units can take advantage of the smallest window of opportunity when it is presented.

## THE MULTIDIMENSIONAL BATTLEFIELD

7-9. The informationized battlefield involves all domains and all dimensions. Commanders must think three dimensionally, and they must effectively integrate capabilities throughout all domains. The PLAA fears enemy airpower, multispectrum surveillance, and precision firepower, and it seeks to offset these capabilities through integrating its different operational systems. At the same time, commanders and planners must ensure that PLA forces can strike through all domains in a synchronized, intelligent manner, to effectively isolate enemy units and defeat advanced enemy systems.

### THE FINANCIAL COST OF MODERN WAR

7-10. The weapons that populate the informationized battlefield are high technology and often very expensive. Outfitting an entire force is an exercise in national economics as much as military strategy. Precision munitions are lethal and effective, but expensive, and stockpiles of them are minimal. The logistics train, which begins with manufacture and ends on the battlefield, must be up to the task of supplying the Chinese force with sufficient modern systems and munitions. Once initial stockpiles are depleted, whichever nation or military that can best sustain its forces will enjoy a significant advantage in combat.

## PRINCIPLES OF THE OFFENSE

7-11. PLAA offensive operations are informed by seven basic principles: concentration, perspective, depth, coordination, adaptation, bravery, and focus. PLAA commanders make maximum use of these principles to multiply the effectiveness of their force, keep the enemy off balance, seize and retain the initiative, and preempt enemy reactions.

### CONCENTRATION

7-12. All PLAA offensive operations seek one essential characteristic: concentration of overwhelming combat power against the enemy at key times and locations. This is a basic and timeless principle, descended from Sun Tzu and Mao, and modified to fit the informationized battlefield. Concentration historically involved maneuver forces and possibly artillery conducting a coordinated action with vast local numerical superiority. Today, however, concentration requires the integration of numerous capabilities across all domains, carefully synchronized and decisively employed. The PLAA now characterizes concentration as an exercise in quality rather than quantity. Instead of focusing on gaining numerical superiority, the command should focus on getting the correct capabilities to achieve the desired effects at the right time and place. Similarly, concentrated forces are seen as potentially vulnerable to enemy counteractions, and they should be carefully guarded by security efforts and counterreconnaissance until their mission is complete. Once complete, forces should either rapidly disperse or swiftly move on to the next target.

### PERSPECTIVE

7-13. The principle of perspective demands two different but interrelated characteristics from PLAA leaders. First, they must be aware of their higher-echelon mission and shape their own decisions based on the objective of their higher echelon. Second, they should seek to know their own strengths, along with enemy weaknesses, to maximize the effectiveness of offensive operations. Offensive objectives should be carefully chosen and offensive capabilities carefully synchronized to create the best possible environment for paralyzing, dismembering, and collapsing the enemy's systems. Once these systems have been defeated, enemy units can be defeated in detail.

### DEPTH

7-14. PLAA doctrine at all echelons places great emphasis on attacks in depth. Depth can refer to the physical—for instance, operating deep behind enemy defenses; the virtual—utilizing offensive cyber or EW capabilities; or the psychological—getting into the enemy's decision cycle, preempting its actions, or breaking its morale. The primary objective of attacks in depth is to isolate enemy formations physically, psychologically, or temporally. Conceptually, the principle of depth is closely related to the ancient axiom of divide and conquer. Enemy units are isolated through envelopment or penetration, weakened, and then annihilated in detail. Attacks in depth keep the enemy off balance, forcing it to fight in numerous locations and preventing it from effectively concentrating forces for an effective counterattack. Depth attacks should be conducted jointly, using all services to create a combined arms effect across the breadth and depth of the enemy formation. The end state is a series of isolated and besieged enemy units and strongholds that can then be annihilated or reduced in detail.

## COORDINATION

7-15. The PLAA considers the informationized battlefield to be incredibly complex, fast moving, and intense. Proactive cooperation between different organizations is a key enabler of offensive actions on the modern battlefield. The PLAA identifies three different components of effective coordination: First, plans should include both a primary plan and numerous backup plans, covering a variety of possible contingencies, and outlining cooperation efforts between different units in each contingency. Second, any changes that take place as situations evolve should be carefully coordinated with adjacent and other concerned units. Third, efforts should be coordinated across the force as a whole, with every subordinate unit conducting operations that support the main effort and commander's intent.

## ADAPTATION

7-16. Adaptability is not a traditional strength of the PLA, but ongoing reforms are trying to change this dynamic in order to better fit the informationized battlefield. Interestingly, while there is an emphasis on growing independence in commanders and a decentralized command approach, the principle of adaptation calls for extensive contingency planning and a scientific approach to unexpected occurrences. Adaptation also emphasizes mobility as a key capability; reaction speed of troops is a major enabler of an adaptive force. Adaptability calls for flexible use of tactics and techniques, and it demands that commanders quickly respond to changing situations on the battlefield with new and creative approaches to the employment of tactics.

## BRAVERY

7-17. The PLAA places a high premium on bravery and aggression in ground actions of all types. Only through audacious and aggressive action will the initiative be gained, and only by gaining and maintaining the initiative can the enemy be annihilated. Though lethality of weapons systems and the complexity of the modern battlefield have changed, the fundamental principle of rapidly closing with the enemy, then aggressively attacking weak points of its formation, remains the basic approach to PLAA offensive tactical actions. The PLAA believes that bravery from its leaders and soldiers leads to decisive engagements, which in turn shorten the length of battles. Bravery in the attack—particularly in the initial assault—helps to take full advantage of the very small windows of opportunity available on the informationized battlefield. One of the most important roles of the political officer is building and maintaining a culture of bravery down to the lowest echelons of the PLAA.

## FOCUS

7-18. The principle of focus has two subordinate concepts. First, it instructs leaders to carefully prioritize their efforts, concentrating their operations against only the most important targets. The PLAA believes that one who prioritizes everything, prioritizes nothing. Second, it demands that commanders properly resource their subordinate units in accordance with the tasks asked of them. Commanders must make their priorities clear to both subordinates and superiors in order to ensure effective prioritization. Subordinates, if given adequate focus by their leaders, are better able to act properly in the absence of direct orders or in other ambiguous situations.

# PLANNING THE OFFENSE

7-19. The PLA historically stressed the importance of careful planning and preparation before an offensive operation. This focus continues in the post-reform PLAA, where offensive actions are developed, rehearsed, revised, and carefully executed whenever possible. Meticulous planning allows for integrating capabilities, developing tight security measures, and building an effective deception plan. While devoting more time for planning may reduce speed in the short term, it is the PLAA's belief that a well-constructed and well- rehearsed plan ultimately saves time and increases operational speed.

7-20. PLAA offensive operations are objective-based: the higher echelon commander designates an objective and specifies a force charged with accomplishing it. People's War principles dictate that objectives should typically be enemy formations, though key terrain or other valuable assets may also be considered. The attacking force identifies key features in the defense and terrain including disposition of enemy forces, weak points and points of strength, locations of reserves and fire support, and key terrain features. The offensive action is then designed to rapidly confuse and isolate the enemy, swiftly reducing its morale and enabling freedom of action for PLAA commanders. Through a combination of careful planning, effective deception, rapid movement, and application of decisive combat power at key times and locations, the enemy comes to believe its position to be untenable and either withdraws or faces annihilation. Once the enemy force is in retreat, reserve forces maintain contact and pursue the enemy while additional forces secure the area and prepare for follow-on operations.

## BUILD THE COMMAND SYSTEM

7-21. An offensive operation makes use of at least two, and ideally four, command posts. The base command post is the commander's primary location, and it should be located to best facilitate coordination between the frontline, depth, and thrust maneuver groups. The advance command post, if established, is led by the deputy commander, and it is typically located near the main defensive line. The rear command post's primary role is to organize logistics and reinforcements and to create the backup defensive line supporting the offensive action. If possible, a reserve command post is established along a possible route of egress or in a well-defended rear location, ready to take over for the base or advance command post should either of them come under threat. At the combined arms battalion (CA-BN) echelon, command post operations may be informal. The command and limited staff at the battalion may fully decentralize and not physically co-locate. Additional information on command post operations can be found in chapter 4.

7-22. The offensive zone is further subdivided into two or more secondary zones, each with a specific set of objectives and different set of tactics. While the PLAA used to be highly prescriptive about the physical sizes of these zones, it has gradually moved to a more flexible approach. The various zones should account for terrain, friendly and enemy capabilities, and higher echelon missions, and they should enable careful integration of various units and capabilities. Figure 7-1 on page 7-7 depicts a combined arms brigade's (CA-BDE's) genericized offensive zone.

### Deep Area

7-23. The deep area is the territory past which a unit's organic sensors and weapons can operate. For a CA-BDE, this typically means the area past which its rocket artillery and targeting support can operate. The fight in the deep area usually consists of independent special operations forces (SOF) or scout units supported by manned or unmanned aircraft (UA), possibly augmented by supporting fires from long-range shooters assigned to support the offensive action. Reconnaissance, counterreconnaissance, fire, counterfire, screening, and blocking all take place in deep areas. The purpose of deep-area operations is to provide early reconnaissance, target long-range preparatory fires, and carefully assess enemy strength and disposition in preparation for an offensive action.

### Frontline Zone

7-24. The frontline zone contains the territory in which the main offensive action is to occur. Early objectives, along with the enemy's main defensive line, are typically located in the frontline zone. The frontline attack group is the primary occupant of this zone, and the depth group may also occupy the area, depending on terrain and enemy disposition. The advance command post, if present, is usually forward in the zone, and the base command post is typically located either rearward in this zone or in the zone immediately behind it. The entire frontline zone should be within the range of the offensive group's organic fire support. The frontline zone typically contains a security zone on its forward edge, where security, reconnaissance, and counterreconnaissance activities take place. The primary battle takes place in the frontline zone, with the intent of breaching the enemy's main defensive line and enabling the depth and thrust maneuver groups to move into enemy rear areas.

## Reserve Zone

7-25. The reserve zone lies just to the rear of the frontline zone and typically houses the depth attack group, thrust maneuvering group, reserve group, command groups, firepower groups, and forward logistics bases. The reserve zone also usually contains a defensive line intended to resist enemy counterattacks into rear areas, and serves as the anchor for the offensive action. The reserve command post, if present, is typically located in this zone, as is the rear command post.

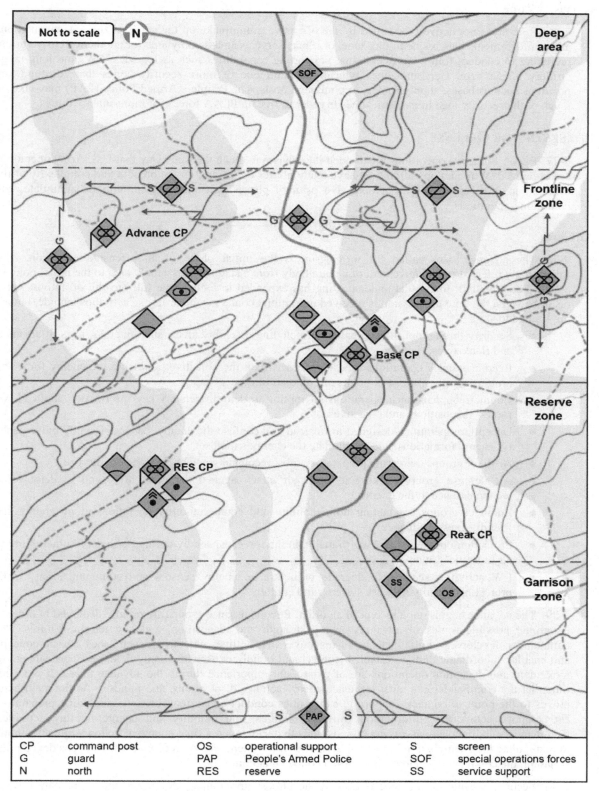

**Figure 7-1. Offensive zone (notional)**

## Garrison Zone

7-26. Rear areas not actively occupied by the offensive group make up the garrison zone. Augmentations and reinforcements may reside in this zone, or it may serve as an assembly area for another offensive group preparing to conduct follow-on operations. Supporting capabilities such as logistics, EW, and long-range artillery reside here. Garrison zones typically contain one or more security zones that surround key positions such as bases, supply routes, or command posts. The People's Armed Police (PAP) may take on much of the security load in garrison zones in order to free up PLAA forces for more-intense duties.

## PHASES OF THE OFFENSE

7-27. PLAA offensive operations are divided into phases in much the same way that U.S. Army operations are organized. While an operation may have many phases depending on the breadth and complexity of the mission, in most cases it will involve five primary phases: advance, unfold, initiate, annihilate, and continuing operations.

## Advance

7-28. The advance, also called the moving-in, is the initial phase of an offensive operation. This traditionally referred to the movement of a main body from a staging or assembly area to the initial point of attack. For a modern force, this understanding has expanded to include the full breadth of activities that occur between the time the mission is received until initial contact is made. These activities include, but are not limited to—

- Security throughout the combat area, including assembly areas, staging areas, axes of advance, and flank areas.
- Reconnaissance operations focusing primarily on the objective: assessing the enemy force and probing for possible weak points or vulnerabilities.
- Counterreconnaissance operations attempting to deny the enemy key information about PLAA forces' dispositions and objectives.
- Deception operations designed to mislead and confuse the enemy, conceal friendly movements, fix enemy formations, and manipulate the enemy's mindset.
- Artillery groups delivering preparatory fires, reconnaissance by fire, and counterfire.
- Air defense groups seeking to deter air attack against the main body and to deny aerial reconnaissance to the enemy.
- Engineer groups conducting both mobility and countermobility operations along enemy and friendly axes of advance.
- Protection operations seeking to maintain the force, especially during periods of vulnerability to air or artillery attack.
- EW activities seeking to degrade or neutralize enemy sensors and communications, while protecting friendly network systems and emitters.

7-29. The advance begins upon receipt of an order. Reconnaissance groups are rapidly deployed in order to determine possible routes of advance, enemy strength and disposition, and key terrain features. The commander develops an initial scheme of maneuver that outlines objectives, establishes a basic concept, and enables subordinate units to establish contact and rapidly close with enemy defenses. Concealment of movement and deception operations are of paramount importance during the advance phase. It is at this time that the commander can most influence enemy actions, dispositions, and mindset. As the main body moves to the point of contact, supporting capabilities conduct concurrent missions, including preparatory fire support, counterfire, information warfare activities, and mobility and countermobility activities. Commanders continuously assess enemy positions to decide upon a final course of action while attempting to manipulate the enemy's mindset and conceal their own intentions. As reconnaissance groups develop the situation, the commander decides upon a final course of action.

7-30. Security during the advance is also of the utmost importance. Deception operations can only be successful if enemy reconnaissance operations are neutralized, spoofed, or defeated by friendly counterreconnaissance. Security elements ensure that main-body movement is unhindered by enemy attack, countermobility efforts, or deception activities. Air defense and protection operations ensure that the main body is not attritted by enemy air attack and artillery during staging or movement.

7-31. The advance phase ends when the main body makes contact with the enemy and the commander initiates actions on contact. Figure 7-2 depicts a PLAA CA-BDE advancing for an attack on an enemy position.

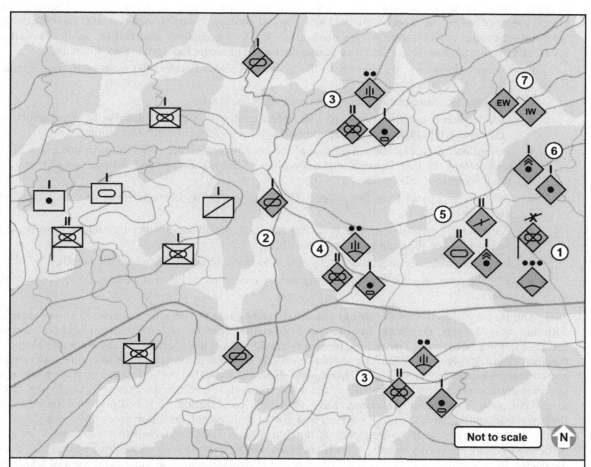

(1) An offensive group advances in preparation for an assault against an enemy mechanized task force that has taken up a defensive position. The offensive group is to annihilate the enemy task force in order to weaken the enemy's overall defensive strength and provide an approach route for follow-on forces. (2) The reconnaissance group deploys in an advance guard position, with orders to reconnoiter enemy defensive positions and identify potential weak points and strong points. (3) Two frontline attack groups comprise the bulk of the main assault. They will fix the enemy and enable actions by the depth attack and thrust maneuver groups. (4) The depth attack group is positioned to exploit any weaknesses in the enemy defenses. This group attempts to conceal its axis of advance in order to surprise the enemy. (5) The thrust maneuver group, consisting of an armor battalion, a mechanized engineer battalion, and a rocket battery, waits in the rear area to exploit the successful attack of the depth attack group. (6) The firepower group, consisting of a heavy howitzer battery and a heavy rocket battery, wait to deliver decisive indirect fire anywhere on the battlefield. (7) IW and EW groups stand by to conduct EW and deception operations, aimed primarily at fooling enemy sensors, deceiving the enemy commander, and suppressing enemy communications.

| EW | electronic warfare | IW | information warfare | N | north |

**Figure 7-2. Advance (example)**

**Unfold**

7-32. The <u>unfold</u> phase consists of actions designed to set conditions for annihilation that occur upon initial contact by the main body with enemy defenses. Subordinate units conduct rapid movement in accordance with the commander's scheme of maneuver, seeking to position themselves appropriately for the decisive phases of the operation. Following initial contact, units maintain contact with the enemy and continue to develop situational understanding. Key intelligence requirements not yet answered by reconnaissance groups are resolved by main-body actions. During initial contact, the commander attempts to conceal the direction, strength, and objective of the main effort through deception, information warfare activities, feints, and demonstrations. Other activities during the unfold phase include, but are not limited to—

- Security operations, particularly to the flanks and rear, to protect the main body from spoiling attacks or counterattacks.
- Reconnaissance operations moving deeper into enemy areas.
- Counterreconnaissance operations that defeat enemy attempts to identify the main effort and objectives while reinforcing deception efforts through feints and demonstrations.
- Deception operations attempting to fool the enemy commander about the time and place of the main effort; the size, strength, and disposition of friendly forces; and the size and proximity of reinforcements for both sides.
- Artillery groups shifting fire support focus to counterfire, obstacle reduction, and disruption efforts in deep areas.
- Air defense groups deterring or defeating aerial attack and surveillance throughout the area, with an emphasis on the main body and main effort.
- Engineer groups shifting focus to obstacle reduction and maintaining friendly mobility.
- EW focusing on jamming or spoofing enemy communications to impede rapid response to the attack.

7-33. The primary objective of the unfold phase is to ensure that the objective is isolated and contending with numerous dangers, preferably from multiple directions. The enemy commander should be confused and enemy forces disrupted. The enemy, however, should not yet know either the direction or strength of the main effort. The entire unfold phase should be orderly and well planned in order to ensure concealment and correct positioning of friendly units. Movement should be rapid, but not so fast that confusion sets in or concealment is broken. The best practice of the unfold phase is to encircle the enemy. Encirclement refers to the complete isolation of the enemy force through a combination of friendly ground unit actions, firepower, deception, psychological warfare, and EW. Encirclement does not necessarily imply that an enemy is actually physically surrounded. A combination of lethal and nonlethal capabilities may be sufficient to cause the enemy to believe itself encircled. For example, a well-timed artillery raid may cause the enemy to believe that a route is cut off or unavailable, without requiring the commitment of additional PLAA ground forces. Similarly, disinformation passed over communications channels may deceive the enemy into believing it does not have access to a particular route or terrain, without requiring friendly forces to physically occupy the area in question.

7-34. The unfold phase ends when the initiate phase begins with the commencement of the main effort. Figure 7-3 depicts a PLAA CA-BDE conducting the unfold phase in preparation for an attack on an enemy position.

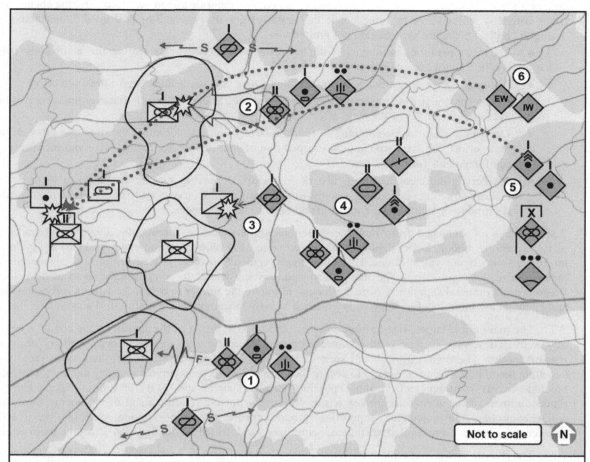

**(1)** One of the frontline attack groups positions in a support-by-fire position, engaging an enemy unit and fixing it in position. **(2)** At the same time, the other frontline attack group conducts a limited attack against the enemy's left flank, testing the strength of the enemy defenses. These two actions, conducted in concert with one another, are intended to confuse the enemy commander about the location and axis of the main assault, forcing him to commit reserves early. **(3)** Reconnaissance units engage enemy scouts, preventing them from effectively reconnoitering friendly units and ensuring the enemy commander remains ignorant of the location and direction of the main assault. **(4)** The depth and thrust groups move under concealment to their initial attack positions. As the frontline attack groups conduct their attacks, the offensive group commander develops his understanding of the situation and finalizes the axis of the main assault. Combat power is concentrated along this axis. **(5)** The firepower group conducts a fire assault against the enemy's rear area, disrupting enemy command and communication and causing casualties. **(6)** EW and IW units conduct electronic and information warfare against the enemy command, with their main effort focused on deceiving the enemy about the location and axis of the main assault.

| EW | electronic warfare | IW | information warfare | S | screen |
|----|--------------------|----|---------------------|---|--------|
| F | fix | N | north | | |

**Figure 7-3. Unfold (example)**

### Initiate

7-35. The <u>initiate</u> phase commences when the main effort begins its assault against the objective. This phase often begins with a massive assault by fire, wherein fire support groups target enemy command nodes, critical systems, mobile reinforcements, and other high-value targets. Supporting efforts continue to develop the situation. Units charged with fixing enemy units continue to prevent them from reinforcing the main objective, while counterreinforcement groups preclude the commitment of enemy reserves. The main effort is typically the primary assault upon the enemy's center of gravity: that thing—be it a headquarters, a unit, or a piece of terrain—that is most essential to the enemy's morale and mindset. Ideally, the enemy will either withdraw, collapse, or be routed once the center of gravity is seized or destroyed. Other activities during the initiate phase include, but are not limited to—

- Security operations continuing with the primary objective of protecting the main body from spoiling attacks or counterattacks.
- Reconnaissance operations focusing on intelligence objectives that support the main effort.
- Counterreconnaissance operations continuing to spoil enemy efforts to discover the composition, axis, and objectives of the main effort.
- Deception operations continuing to fool the enemy commander about the time and place of the main effort; the size, strength, and disposition of friendly forces; and the size and proximity of reinforcements for both sides.
- Artillery groups conducting a firepower assault, delivering massed fire support to the main effort, targeting defending enemy units, interdicting movement of reinforcements, and destroying or suppressing command and communication nodes.

7-36. Attacks are launched from multiple directions simultaneously, forcing the enemy commander to make decisions about force deployment. If possible, the enemy commander's decision cycle is interrupted—or, better yet, influenced—so that reinforcements or other maneuvers do not concentrate in opposition to the main effort. The main effort employs substantial force concentration to achieve local firepower and numeral superiority over the enemy, allowing the former to defeat the latter's forces in detail and maintain the initiative. Fire support is weighted heavily in support of the main effort's assault.

7-37. The main effort targets one or more primary objectives as a part of the assault on the enemy's center of gravity. These may include enemy command posts, assembly areas, reserve forces, artillery units, or network nodes. The main effort should seek weak points in the enemy's defense, then move rapidly to penetrate them. Follow-on forces help to secure the main effort, fix or deceive remaining enemy forces, and preclude enemy reinforcement. Groups—possibly airborne, air assault, SOF, or militia—located to the flanks and rear of the enemy secure key terrain to prevent outside reinforcement of the enemy formation while encircling the objective. Supporting groups ensure that enemy counterattacks—particularly counterencirclement—are spoiled, disrupted, neutralized, or defeated before they can affect the main effort.

7-38. The initiate phase is complete once the enemy center of gravity has been seized by the main effort. Figure 7-4 depicts a PLAA CA-BDE commencing the initiate phase against an enemy position.

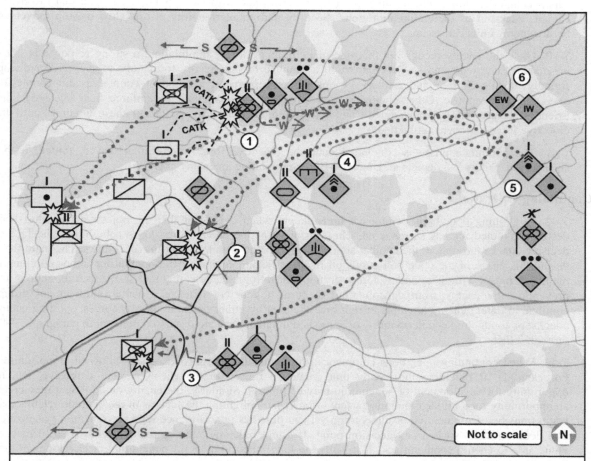

**(1)** The enemy commander, deceived into thinking that the main effort is targeting his left flank, commits his reserve in a counterattack against the northern frontline attack group. The latter rapidly transitions to a defensive posture and begins a deliberate retrograde operation, intended to overextend the enemy's counterattack force and eventually isolate it from its command. **(2)** The friendly commander, having identified the enemy's center as vulnerable, commits the depth attack group in a decisive assault. Its objective is to breach the enemy's main defensive line and isolate the two enemy flanks. **(3)** The other frontline attack group conducts an assault against the enemy's right flank, fixing the enemy unit and preventing it from reinforcing the group under assault in the center. **(4)** The thrust group positions itself to exploit the breach created by the depth attack group. **(5)** The firepower group conducts a fire assault against the enemy center in support of the depth group's assault. **(6)** The EW group commences its decisive effort, suppressing enemy communications in order to electronically isolate enemy units and confuse response to the main assault. The IW group transitions to information attack operations, attempting to increase enemy units' perception of isolation.

| | | | | | | |
|---|---|---|---|---|---|---|
| B | breach | F | fix | N | north |
| CATK | counterattack | IW | information warfare | S | screen |
| EW | electronic warfare | | | | |

**Figure 7-4. Initiate (example)**

## Annihilate

7-39. The <u>annihilate</u> phase commences after the enemy is confused and demoralized following the seizure or fall of its center of gravity. Enemy units should be isolated, with communications to their higher echelons disrupted and routes for reinforcement either cut off or unusable. The enemy is encircled, either physically or psychologically, and mutual support between enemy units is no longer possible. Decisive attacks are conducted throughout the combat area, seeking to destroy isolated enemy units before retreat and a prepared defense are possible. Fire support is used aggressively in brief but violent volleys in order to suppress and demoralize enemy units prior to assaults. Special focus is given to ensuring that the enemy cannot break out of the annihilation zone. Security and countermobility efforts slow or stop movement before enemy units can escape.

7-40. Once the enemy recognizes that one of its units is facing annihilation, a strong response is likely inevitable. Enemy air and artillery units will attempt to target friendly formations to suppress or disrupt friendly attacks. Antiair and artillery groups must be prepared to conduct counterair and counterfire operations in response. Other enemy ground formations will attempt to reinforce their besieged units. Friendly security forces must defend stubbornly and disrupt, slow, or defeat these attempts until the objective force has been annihilated.

7-41. The annihilate phase is concluded once the objective force has been destroyed, routed, or has surrendered. Figure 7-5 depicts a PLAA CA-BDE annihilating an enemy position.

## Continuing Operations

7-42. Continuing operations commence immediately when it is clear that the objective force has been destroyed, routed, or has surrendered. The remainder of the enemy force loses cohesion in the face of decisive attacks throughout the combat area, and widespread withdrawal or retreat follows. The main focus during this phase of the operation is pursuit of these fleeing enemies. High-tempo operations continue to maintain contact with these enemy forces during their retrograde, defeating rear-guard security actions and preventing the enemy from consolidating to form a coherent defense. Counterreinforcement groups continue to defeat enemy attempts to reinforce the area by either defeating enemy forces, spoiling enemy attacks, or conducting countermobility operations. Pursuit actions are discussed in detail in paragraphs 7-93 through 7-96.

7-43. When continuing operations commence, the commander begins assessing follow-on opportunities. For example, if defeating the enemy unit created a breach in the enemy's main defensive line, an opportunity for envelopment or penetration may be present. Similarly, the enemy unit's defeat may create an opportunity for a Public Opinion Warfare offensive, highlighting the success of the operation. Units may begin transition from combat postures to movement postures in accordance with the scheme of maneuver. The decision to do so is a key decision point. If a unit redeploys into march order too soon, it may be vulnerable to enemy counterattack. If it waits too long, opportunities for penetration or envelopment may be lost.

7-44. The PLAA unit must also consolidate. Consolidation consists of three primary activities: perform security activities, reorganize and reconstitute friendly units, and conduct passage of lines. Security activities consist of securing the combat area against enemy reconnaissance and counterattack and ensuring that friendly forces entering the area are not subject to threat from bypassed enemy elements or irregular forces. Reorganization and reconstitution are those activities that enable the unit to conduct follow-on operations: resupply, replacement of lost or damaged equipment, casualty processing, and receipt of reinforcements. Passage of lines takes place when one unit passes through another's combat area, typically with the intent to resume offensive operations against the enemy. Passage of lines can be a complex task, and it is enabled by effective control measures, communications, and planning.

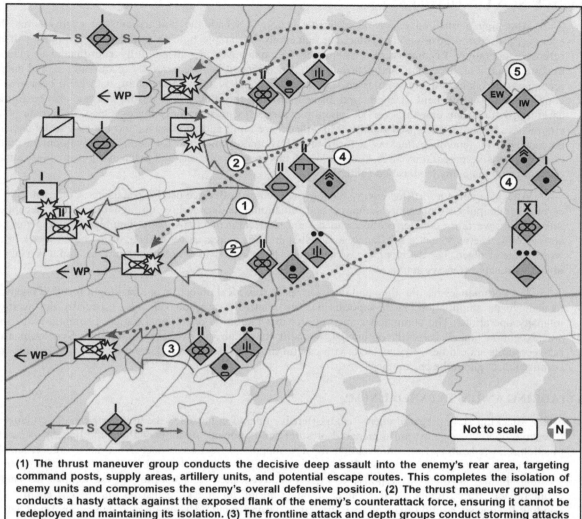

(1) The thrust maneuver group conducts the decisive deep assault into the enemy's rear area, targeting command posts, supply areas, artillery units, and potential escape routes. This completes the isolation of enemy units and compromises the enemy's overall defensive position. (2) The thrust maneuver group also conducts a hasty attack against the exposed flank of the enemy's counterattack force, ensuring it cannot be redeployed and maintaining its isolation. (3) The frontline attack and depth groups conduct storming attacks against the isolated and depleted enemy units. Local fire support is integrated with maneuver to either destroy or force the withdrawal of enemy units. Assaults are coordinated as much as possible to ensure the enemy cannot reinforce units under attack. (4) Firepower assaults target retreating enemy units, disrupting their movement and ensuring that retrograde actions cannot be mounted. (5) EW and IW groups shift their focus to disrupting adjacent enemy units from reinforcing the now-defeated enemy unit and preventing the enemy's higher echelon from communicating with the defeated unit.

| EW | electronic warfare | N | north | WP | withdrawal under pressure |
|----|--------------------|---|-------|-----|---------------------------|
| IW | information warfare | S | screen | | |

**Figure 7-5. Annihilate (example)**

# TYPES OF OFFENSIVE OPERATIONS

7-45. The PLAA recognizes at least five different types of offensive operations: attacking a fortified enemy, attacking an unprepared enemy, amphibious operations, urban offensive operations, and airborne operations. Though they are broadly similar to deliberate and hasty attacks, the primary difference between attacking a fortified enemy and attacking an unprepared enemy is the enemy's level of preparation in contrast to the friendly forces' level of preparation. Other offensive operations focus on specific situations—amphibious, airborne, or urban. In addition to the five types of offensive operations, particular attention is given to operations taking place in special conditions, such as jungles or extreme cold.

## ATTACKING A FORTIFIED ENEMY

7-46. Attacking a fortified enemy is offensive action in which the unit assaults an enemy force that occupies a prepared position. The attacking unit likely has ample time to conduct all five phases of the offensive operation. Reconnaissance, firepower, and maneuver are all carefully coordinated to achieve maximum surprise, deception, and concentration. Enemy positions are carefully reconnoitered, and enemy reconnaissance is effectively suppressed. The advance phase is done deliberately, with great care used to conceal friendly movement, strength, and disposition from the enemy. The unfold phase is also performed carefully, with strong emphasis on deception and concealment of friendly forces. Once the initiate phase commences, tempo is increased rapidly. Movements are rapid and aggressive, and the unit seeks decisive contact with the enemy. Attacking a fortified enemy may allow time for complete envelopment of the enemy, extensive preparatory fires, positioning of forces in deep areas, and extensive information warfare operations upon enemy leaders and soldiers.

7-47. Attacks against a fortified enemy must make the most effective use of stratagems. Commanders can carefully manipulate the enemy's morale and situational understanding through the use of deception, information warfare, counterreconnaissance, and concealment. Effectively employed stratagems can convince the enemy to cease resistance, to conduct an early retreat or retrograde, or to counterattack at the wrong time or place. The development of stratagems begins early in the planning process, and it is a central objective in any offensive action.

7-48. If a commander has the time and decision space, the ideal objective in attacking a fortified enemy is to achieve its annihilation. Such opportunities are relatively rare and valuable, especially during high-intensity operations. The commander must ensure that the unit's operational tempo, planning, rehearsals, and execution are all carefully optimized, but still fit within the available timeframe. Synchronization between units should be well executed, with careful timelines designed to keep the enemy off balance and confused throughout the engagement.

## ATTACKING AN UNPREPARED ENEMY

7-49. Attacking an unprepared enemy is an offensive operation in which the unit must rapidly conduct an offensive operation against an unprepared or surprised opponent, possibly without the benefit of careful positioning, preparatory fires, information warfare support, or reconnaissance. Such an operation may be one of opportunity—such as a friendly unit discovering a previously unseen weakness in an enemy's position. It may also be in response to enemy actions, such as a spoiling action against an enemy counterattack. Rapid tempo is of great importance during an attack against an unprepared enemy. The window for decisive action may be very limited, requiring rapid decision making and action on behalf of the attacking unit.

7-50. Unprepared enemies are typically vulnerable, or they hold disadvantageous positions. Careful reconnaissance may not be possible on a limited timeline, so attacking an enemy in a strong defensive position without proper preparation should be discouraged. The rapidity and aggressiveness of the attack is the key enabler to the mission. It keeps the enemy off balance and grants the initiative to the friendly unit commander.

7-51. Attacks against an unprepared enemy still include all five phases of an offensive operation, but one or more phases may be very short in duration or limited in scope. The advance phase is short, with units forgoing careful movement, deception, and concealment for rapidity—they seek to close with the enemy as quickly as possible, ensuring it cannot construct a strong defensive position. The unfold phase is conducted in a similar manner, with rapidity of action being prioritized over deception and concealment. While in an attack against a fortified enemy, units may be positioned on a number of different approach axes. An attack against an unprepared enemy may allow for as few as two. From the initiate phase onward, operations are broadly similar to those in an attack against a fortified enemy. The enemy is struck with speed and aggression, and the attacking units seek to isolate, overwhelm, and then annihilate enemy units in detail.

*Note.* Chinese military theory suggests no attack be made with fewer than two axes of advance against a single point.

7-52. A unit conducting this type of attack is particularly vulnerable to enemy counterattack. The unit likely did not have the time or resources to deploy a comprehensive security element, and it may be deployed in a vulnerable or disorganized fashion due to the rapid push through the unfold phase of the attack. Friendly commanders must be prepared to react to a well-timed enemy counterattack, either through fire support, information warfare, or direct action by ground or air forces. Developing and maintaining security following the attack is a critical component of the continuing operations phase.

7-53. Successful attacks against unprepared opponents require a significant degree of autonomy by the attacking units. The small time window for success and the lack of careful preparation often preclude careful coordination with higher echelons. Enabling this level of autonomy is a significant focus area for the PLAA. Historically, it was rare for units above the battalion echelon to conduct any such actions. Modern PLAA CA-BDE and CA-BNs are expected to be able to conduct quick attacks with only limited coordination and support from their group army.

## AMPHIBIOUS LANDING OPERATIONS

7-54. The PLA defines amphibious landing operations as those that make use of watercraft to cross a body of water, enabling the attacking unit to conduct an assault and exploitation against an enemy defending a littoral area or island. The PLAA recognizes two distinct kinds of amphibious landing operations: coast to coast and vessel to coast. Coast-to-coast operations are those in which staging and embarkation occur on shore, while vessel-to-coast operations stage and embark from a larger seagoing vessel.

7-55. Amphibious landing operations use a similar approach to operations across land. The largest differences are that the advance phase occurs across water and reconnaissance efforts are likely to be less comprehensive. An amphibious landing operation consists of an advance phase, an assault and penetration phase, and an exploitation and consolidation phase. Secrecy and deception during embarkation and advancing are considered paramount, as embarked troops are highly vulnerable to counterattack and firepower attack. Ideally, the initial wave of an amphibious assault strikes an undefended or lightly defended beach, allowing the first wave to establish a beachhead security area while under minimal enemy fire. Once the beachhead is established, follow-on waves land within the secure area, then conduct penetrations of deeper enemy positions. If at all possible, multiple beachheads should be established, enabling multiple follow-on penetrations and dispersing the effects of enemy counterattack and enabling.

7-56. Amphibious landing operations are to be supported by massive firepower and information attack. Firepower may be generated from missiles, aircraft, surface vessels, or long-range artillery. Information attacks maximize the effect of deception operations and seek to extend enemy response times. Once ashore, the initial wave establishes substantial firepower superiority over local enemy defenses, forcing enemy retreat and enabling the establishment of a beach security area. Reconnaissance likely occurs concurrently with the assault. Reconnaissance elements should deploy and conduct reconnaissance in force to ascertain enemy intent and dispositions and inform the command about how and where to commit penetration efforts.

7-57. The decision of where and how to commit the follow-on efforts is the most important decision in an amphibious landing operation. Commanders in the initial wave must communicate their situational understanding—enabled by on-the-spot reconnaissance efforts—to their higher echelon, allowing the higher echelon to decide how and where the second wave should attack. Commanders in the first wave must also coordinate firepower operations to support follow-on waves and clear any remaining obstacles or other enemy countermobility assets to enable the penetration effort. Figures 7-6 through 7-8 on pages 7- 18 through 7-20 show an amphibious operation.

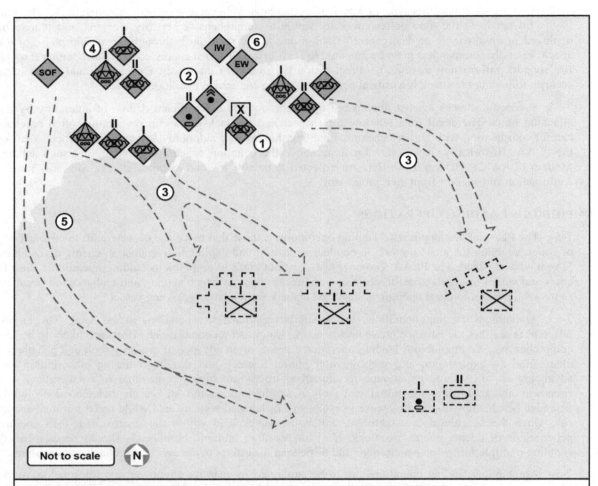

(1) An amphibious group built around a PLANMC brigade conducts an amphibious assault on a contested enemy shoreline. The group is to establish a beachhead, drive enemy forces from the area, and secure the area to enable follow-on operations. (2) Long-range fire support for the operation is provided by a heavy MRL battery, while a SPG battalion prepares to move across the strait once the beachhead is established. (3) Two frontline attack groups, each consisting of an amphibious reconnaissance company, amphibious infantry battalion, and amphibious tank/assault gun company, are to transit the strait and engage enemy shore defenses. (4) A depth attack group stands by to exploit the weakest part of the enemy's defenses and strike inland. (5) A SOF company is to conduct an air insertion into the enemy's depth area. (6) EW and IW groups begin EW and deception operations, attempting to suppress enemy sensors and communications and fool the enemy about the time and place of the main effort's assault.

| EW | electronic warfare | PLANMC | People's Liberation Army Navy Marine Corps |
|-----|------------------|--------|-------------------------------------------|
| IW | information warfare | S | screen |
| MRL | multiple rocket launcher | SOF | special operations forces |
| N | north | SPG | self-propelled gun |

Figure 7-6. Amphibious landing operation (example, part 1 of 3)

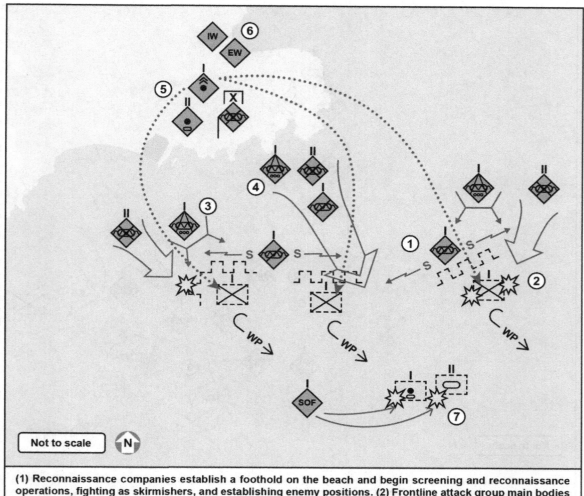

(1) Reconnaissance companies establish a foothold on the beach and begin screening and reconnaissance operations, fighting as skirmishers, and establishing enemy positions. (2) Frontline attack group main bodies conduct assaults against enemy weak points, seeking to penetrate enemy shore defenses and fix enemy units. (3) Frontline attack group assault guns provide support by fire both from shore and while embarked. (4) The depth attack group conducts a decisive assault through the enemy center, forcing an enemy retreat and isolating remaining enemy defenders. (5) The firepower group delivers long-range fire support at key points on the battlefield. (6) EW and IW groups continue to dominate the electromagnetic spectrum and deceive the enemy about the time and place of the main assault until the last possible moment. (7) The SOF company conducts raids in depth areas, disrupting enemy reinforcement and fire-support missions.

| EW | electronic warfare | N | north | SOF | special operations forces |
|----|--------------------|---|-------|-----|---------------------------|
| IW | information warfare | S | screen | WP | withdrawal under pressure |

**Figure 7-7. Amphibious landing operation (example, part 2 of 3)**

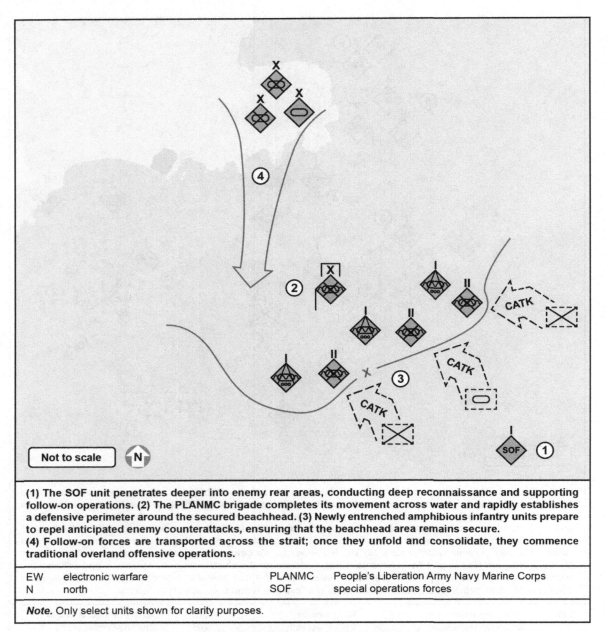

(1) The SOF unit penetrates deeper into enemy rear areas, conducting deep reconnaissance and supporting follow-on operations. (2) The PLANMC brigade completes its movement across water and rapidly establishes a defensive perimeter around the secured beachhead. (3) Newly entrenched amphibious infantry units prepare to repel anticipated enemy counterattacks, ensuring that the beachhead area remains secure.
(4) Follow-on forces are transported across the strait; once they unfold and consolidate, they commence traditional overland offensive operations.

| EW | electronic warfare | PLANMC | People's Liberation Army Navy Marine Corps |
| N | north | SOF | special operations forces |

*Note.* Only select units shown for clarity purposes.

**Figure 7-8. Amphibious landing operation (example, part 3 of 3)**

## URBAN OFFENSIVE OPERATIONS

7-58. Urban offensive operations are those offensive operations that take place in complex urban terrain. In contrast to maneuver warfare in more open terrain, the PLAA stresses careful planning and execution, slow and deliberate actions, and meticulous intelligence assessments when conducting offensive operations in urban areas.

7-59. Urban offensive operations are characterized by bounding overwatch techniques, using streets or other thoroughfares as avenues of approach. Following an extended period of reconnaissance across the area, a forward element attacks and secures a position of overwatch. Ideally, this element occupies an elevated position in order to provide fire and reconnaissance support to assault elements. Assault elements then move down avenues of approach while targeting enemy weak areas. Urban operations are viewed as fundamentally slower and more deliberate than other operations for both sides, Surprise and deception are even more important, as actions take longer and attacking units are more vulnerable.

7-60. Fighting in urban areas is also characterized by increased complexity, particularly of terrain. High-rise buildings are areas of special focus. If a large building must be taken, it should be carefully cordoned, isolated, and then aggressively stormed by specially trained troops. Underground facilities are defeated in a similar manner. The employment of specialized troops to deal with these difficult targets is a critical enabler of urban offensive operations.

7-61. Complex urban terrain also changes the nature of the enemy's counterattack. Buildings and other hardened pieces of terrain make concealed movement simpler for the enemy's counterattack force, but they also enable and enhance ambush operations. PLAA commanders should seek to defeat or demoralize enemy counterattack forces through ambushes that use the urban terrain. Security units should defend stubbornly and protect the main body as it deliberately and thoroughly defeats any lingering enemy units within the combat area. In addition, urban terrain enhances the use of countermobility capabilities, particularly in tight, canalized areas. Commanders should liberally employ countermobility systems in careful concert with ambush and security operations. Figure 7-9 on page 7-22 depicts a PLAA CA-BN conducting an urban offensive operation.

## AIRBORNE OPERATIONS

7-62. Airborne operations are offensive operations using fixed- or rotary-wing aircraft to move friendly troops into enemy territory. Following airborne insertion, units conduct otherwise typical maneuver warfare. Airborne operations are one of the most important elements of the multidimensional penetration, the use of all domains and dimensions to create multiple dilemmas for the enemy. Currently there are relatively few PLAA airborne transportation assets, and airborne units are considered high-value assets. They are prescribed to target the highest-value enemy assets within their operating area: the flanks and rear of the enemy main body, airfields, artillery, air defense systems, or key routes for enemy reinforcements.

7-63. Airborne operations are broadly similar to amphibious operations, though they are conducted on a smaller scale. After careful but rapid reconnaissance and a firepower attack, an assault team attacks and seizes the landing area, annihilating enemy forces in the vicinity and creating a secure base from which follow-on operations can be launched. Additional forces are airlifted to the secure base, and they conduct aggressive offensive operations against key targets in their areas of operations. Security operations are ongoing, seeking to spoil or defeat enemy counterattacks.

7-64. Airborne forces may be relieved in place by ground forces, or they may be exfiltrated by air. Both processes are challenging. Relief in place requires careful coordination between friendly units to avoid both friendly fire and excessive vulnerability to enemy attack, while air exfiltration must be conducted rapidly and efficiently to minimize the risk to friendly aircraft.

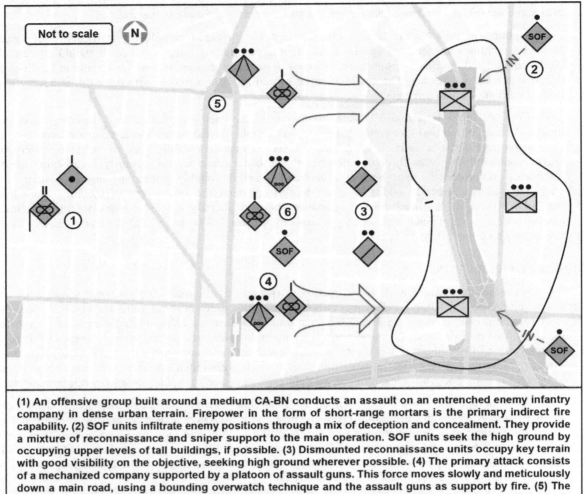

(1) An offensive group built around a medium CA-BN conducts an assault on an entrenched enemy infantry company in dense urban terrain. Firepower in the form of short-range mortars is the primary indirect fire capability. (2) SOF units infiltrate enemy positions through a mix of deception and concealment. They provide a mixture of reconnaissance and sniper support to the main operation. SOF units seek the high ground by occupying upper levels of tall buildings, if possible. (3) Dismounted reconnaissance units occupy key terrain with good visibility on the objective, seeking high ground wherever possible. (4) The primary attack consists of a mechanized company supported by a platoon of assault guns. This force moves slowly and meticulously down a main road, using a bounding overwatch technique and the assault guns as support by fire. (5) The second assault group times its attack to commence once the enemy is committed to repulsing the first attack. It must move more quickly but meticulously down an adjacent avenue of approach. (6) The depth group provides support by fire to both assaults, and will conduct its own assault if required.

| | | | |
|---|---|---|---|
| CA-BN | combined arms battalion | N | north |
| IN | infiltrate | SOF | special operations forces |

**Figure 7-9. Urban offensive operation (example)**

OPERATIONS UNDER SPECIAL CONDITIONS

7-65. The PLAA recognizes eight special conditions: hills, mountains, jungles, extreme cold, deserts, rivers, swamps, and nighttime. Each of these conditions requires modifications and adaptations to baseline operations. Primary considerations include ensuring that troops are properly prepared for the special condition-with cold weather gear or bridging equipment, for example. Units should also train for potential special conditions that may occur in their areas of operations.

# OFFENSIVE TACTICS

7-66. The PLAA employs several different offensive tactics: envelopment, penetration, pursuit, firepower, ambush, and raid. Use of these tactics is informed by the principles discussed earlier in this chapter. Each tactic may be employed by virtually any echelon, but the figures in this chapter show employment at the CA-BDE level against a notional enemy. Tactics may be used independently, sequentially, or simultaneously. Figures in this section depict genericized units conducting simplified variants of the tactics described.

## ENVELOPMENT TACTICS

7-67. Two primary envelopment methods are used by the PLAA: simple envelopment and complex envelopment—both of which occur on enemy flanks. These methods are described in paragraphs 7-68 through 7-75.

### Complex Envelopment

7-68. A complex envelopment is an offensive tactic wherein an assault occurs against multiple flanks of the enemy. The complex envelopment has long been a mainstay of Chinese maneuver tactics, employed regularly by Chinese generals throughout the various wars of Imperial China. More recently, it was used regularly—and with good effect—by Communist Chinese forces in both the Chinese Civil War and the Korean War, most famously at the 1946 Battle of Guanzhong and the 1951 Battle of the Soyang River. A successful complex envelopment creates multiple dilemmas for the enemy commander, enables effective isolation of enemy units, and enables concentrated forces to defeat enemy strongpoints in detail. For a contemporary maneuver unit, a complex envelopment requires advanced coordination of multiple maneuvers along with effective integration of fires, information warfare, and mobility and countermobility capabilities.

7-69. The definition of flank traditionally refers to the right or left side of a ground formation. The PLAA recognizes two distinct types of complex envelopment: the flank envelopment, wherein attacks target the enemy flanks, and the separating envelopment, wherein attacks target enemy positions of weakness in depth. The PLAA's multi-domain approach to fighting recognizes that "flanks" can exist not only on the sides of an enemy formation, but in the air, below ground, or in the virtual world. A modern complex envelopment may make use of any or all of these in a coordinated, synchronized manner in order to get the enemy off balance, preclude effective enemy response, gain and maintain the initiative, and get inside the enemy's decision cycle. In addition, a penetration attack may create flanks where they did not previously exist; a network attack may threaten or disrupt the enemy's digital or virtual flank; and air or missile forces may threaten an enemy's flank in the third dimension. A flank, in other words, is defined by the PLA as any area in which the enemy has not concentrated defenses, does not expect an attack, or is vulnerable. In general, the more flanks are threatened or engaged simultaneously, the less coherent and effective the enemy's response will be. This in turn enables isolation of enemy units and eventually leads to their destruction.

7-70. The complex envelopment employs one or more depth attack groups enabled by one or more frontline attack groups. A frontline attack group attacks, feints, or demonstrates against one portion of the enemy formation to fix or deceive enemy forces, engage enemy artillery, and drive commitment of enemy reserves. Once the enemy has been fixed, the depth attack groups conduct aggressive, penetrating assaults against weak points of the enemy's flank, while reserve and artillery groups continue to fix other enemy forces, preventing effective reaction to the penetration. Depth attack groups seek to disrupt enemy rear areas and isolate enemy units, enabling follow-on annihilation attacks. Depth attack groups must achieve substantial overmatch against their initial opposition, integrating numerical superiority, firepower superiority, and information superiority to overwhelm the enemy and enable rapid maneuver into and through enemy rear areas.

7-71. Success of a complex envelopment depends heavily on deceiving the enemy about the strength and axis of the depth attack groups. Deception operations and careful concealment enable the assault forces to strike from an unanticipated direction against areas of weakness. The operations of the frontline attack groups prevent the enemy from counterattacking or reinforcing the units opposing the depth assault until it is too late to adequately respond. Figure 7-10 on page 7-24 depicts a complex envelopment.

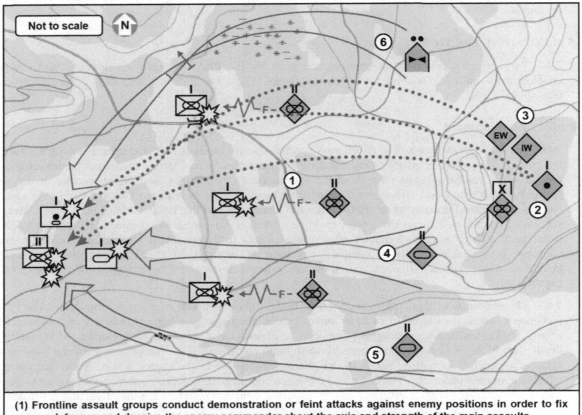

(1) Frontline assault groups conduct demonstration or feint attacks against enemy positions in order to fix enemy defenses and deceive the enemy commander about the axis and strength of the main assaults.
(2) The firepower group delivers indirect fire against enemy artillery and command elements in order to disrupt enemy forces and suppress enemy fires. (3) EW and IW groups conduct an attack against the enemy's digital flank, exploiting network vulnerabilities, suppressing communications, and amplifying the effects of deception efforts. (4) The depth attack group conducts a separating envelopment against the enemy center in order to isolate forward enemy units and force the enemy to commit reserves. (5) The depth attack group conducts a flank envelopment, penetrating into the enemy rear and threatening the exposed enemy flank. (6) The depth attack group conducts a flank attack by air assault into the enemy rear area, cutting off retreat and completing isolation of forward enemy units.

| EW | combined arms battalion | IW | infiltrate |
|----|-------------------------|----|------------|
| F | fix | N | north |

**Figure 7-10. Complex envelopment (example)**

## Simple Envelopment

7-72. The simple envelopment is an offensive tactic wherein an assault occurs on a single flank of the enemy. A simple envelopment involves two primary groups: a depth attack group and a frontline attack group. The frontline attack group feints or demonstrates against one portion of the enemy formation, with the intent of fixing enemy maneuver forces, engaging enemy artillery, and driving enemy commitment of reserves. Frontline attack group actions can be thought of as a deception operation, intending to prevent the enemy from concentrating its forces against the depth attack group.

7-73. Once the enemy is engaged by the frontline attack group, the depth attack group advances to its attack position and commences an assault. The depth attack group seeks to achieve vastly superior concentration of combat power against the specified flank of the enemy. Maneuver forces, enabled by integrated fires and information warfare support, overwhelm any prepared defenses, and penetrate enemy rear areas. Once the enemy's flank is defeated and its rear areas are threatened, isolated enemy units can be annihilated by follow-on operations.

7-74. Much like with a complex envelopment, success of a simple envelopment depends heavily on deceiving the enemy about the strength and axis of the depth attack group. Deception operations and careful concealment enable the assault force to strike from an unanticipated direction against areas of weakness. The operations of the frontline attack group prevent the enemy from counterattacking or reinforcing the units opposing the depth attack group until it is too late to adequately respond.

7-75. The simple envelopment is employed in situations where a complex envelopment is not possible due to terrain, strength, or disposition. Use of only a single penetration makes it easier for the enemy to respond and effectively counter the envelopment attempt. If possible, the enemy should face multiple penetrations to complicate and confuse responses. Simple envelopments are very common at the squad and platoon levels, where one element creates a base of fire to fix and suppress the enemy, while another element assaults the objective. Figure 7-11 depicts a simple envelopment.

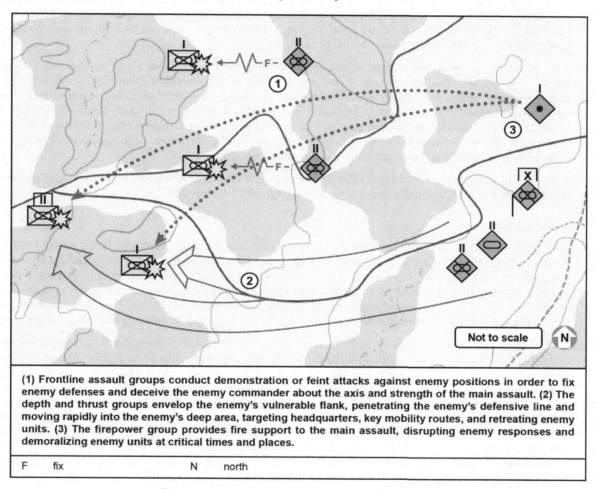

(1) Frontline assault groups conduct demonstration or feint attacks against enemy positions in order to fix enemy defenses and deceive the enemy commander about the axis and strength of the main assault. (2) The depth and thrust groups envelop the enemy's vulnerable flank, penetrating the enemy's defensive line and moving rapidly into the enemy's deep area, targeting headquarters, key mobility routes, and retreating enemy units. (3) The firepower group provides fire support to the main assault, disrupting enemy responses and demoralizing enemy units at critical times and places.

| F | fix | N | north |
|---|-----|---|-------|

**Figure 7-11. Simple envelopment (example)**

## PENETRATION

7-76. A penetration is an offensive tactic wherein a mobile or fast-moving element defeats an enemy line of defense—preferably through an area of weakness—and continues to press forward into the enemy's rear area, bypassing enemy areas of strength. A penetration is rarely employed by itself. Instead, it is typically employed as one component of a wider operation, such as an envelopment. Ideally, an offensive campaign successfully conducts numerous multidimensional penetrations, employing a variety of capabilities to penetrate enemy defenses in a synchronized and mutually supporting manner. Units best suited to conduct a penetration as part of a ground action are armored or mechanized units due to their combination of mobility, protection, and firepower. Penetrations may also be employed in the air or cyber domains or in the electromagnetic spectrum using the same basic principles: target enemy weak points and bypass strong points in order to threaten vulnerable areas behind the main lines of defense.

7-77. Most contemporary PLAA tactics involve one or more penetrations as the basis for offensive operations. This is true for all echelons and unit types. Penetrations are the means by which enemy units are isolated, enemy actions and counterattacks are disrupted, and additional enemy flanks are exposed. Aggression and mobility are the key enablers of a successful penetration, along with effective integration of enablers that preempt enemy response to a penetration, deceive the enemy as to the axis or strength of a penetrating force, or create confusion or the perception of isolation among enemy units during a penetration.

7-78. As it requires a unit to bypass active enemy positions, a penetration by itself leaves the conducting unit highly vulnerable to counterattack on its flanks and rear. In order to maximize the effect of a penetration, the main effort must be well supported by security operations and follow-on operations designed to secure the areas bypassed by the penetration.

7-79. Similarly, once a penetration is achieved, it must be exploited in order to maximize effect. Exploitation can take a variety of forms. For a ground action, it may consist of follow-on operations against enemy rear area assets, annihilation operations against isolated enemy units, security operations that offset enemy reconnaissance and counterattack efforts, or consolidation efforts centered on newly seized areas or terrain. Exploitation in the air domain may include ground attack or strike operations; in the cyber domain, anything from passive observation to network attack; and in the electromagnetic spectrum, jamming or meaconing (the interception and surreptitious rebroadcasting of signals).

7-80. The PLAA practices four types of penetration: frontal attack, depth attack, infiltration, and storming attack. These types of penetration are described in paragraphs 7-81 through 7-92.

### Frontal Attack

7-81. A frontal attack is an offensive tactic wherein a penetration is attempted directly against an enemy's position of strength. This tactic is generally not preferred, as it allows the enemy to make best use of its strengths, making effective concentration and the development of overwhelming combat power more difficult. Frontal attacks should only be conducted either to enable offensive actions or because no other alternative is available, but an offensive action is necessary. Frontal attacks need not involve maneuver forces; any offensive action that attacks an enemy's area of strength can be considered a frontal attack.

7-82. The frontal attack is opposed to both Sun Tzu and Maoist theories about warfare, which prescribe attacking the enemy at weak points. In a frontal attack, friendly forces should be highly concentrated and should close as rapidly as possible with the enemy in order to minimize exposure to defensive measures and stand-off weapons. If possible, a frontal attack should be carefully planned and executed to maximize the use of enabling capabilities. Any frontal attack should be enabled by massed fire, substantial information warfare and deception support, and feints or demonstrations designed to preempt enemy actions, confuse and demoralize the enemy, or suppress the enemy. Enabling capabilities should be carefully synchronized with the attack to maximize their effectiveness at key times and places. Even with these enablers effectively used, a frontal attack will likely face significant resistance and will suffer higher casualties than an attack targeted at an enemy weak point. Figure 7-12 depicts a frontal attack.

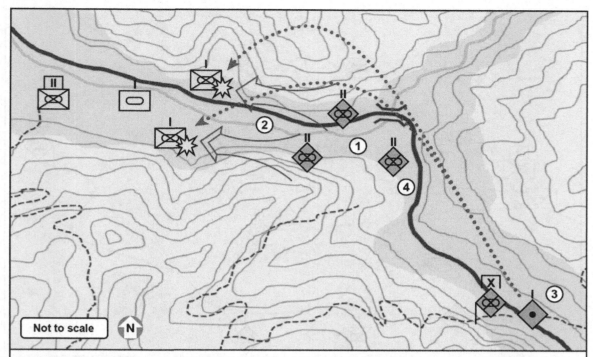

(1) An offensive group is canalized by terrain but must make an assault against a strong enemy position. The commander orders a frontal attack against the entrenched enemy. Two frontline attack groups are massed against an outnumbered enemy. (2) The primary assault concentrates a full battalion in an attempt to penetrate and compromise the enemy's main defensive line. A supporting attack is synchronized in order to prevent the enemy from massing greater combat power against the main effort. (3) Firepower is massed in support of the frontal attacks, attempting to disrupt enemy defenses as much as possible. (4) A depth group awaits successful penetration of the enemy's main defensive line, prepared to exploit the advantage and assault the enemy's rear area.

N     north

**Figure 7-12. Frontal attack (example)**

**Depth Attack**

7-83. A depth attack is an offensive tactic wherein a force focuses on attacking targets deep behind the enemy's forward positions through a series of deep penetrations. A depth attack targets enemy high-value assets in enemy rear areas, such as command nodes, supply bases, reserve troop concentrations, and artillery units. The depth attack also seeks to further isolate enemy units throughout the battlefield, allowing their encirclement and destruction in detail.

7-84. The depth attack is typically the main effort of a given battle plan, and it often involves the best-equipped and -trained troops available to a commander. The depth attack group is typically employed as the second line of troops, after the initial line engages the enemy and creates one or more breaches in the enemy's defensive line. The timing and direction of the depth attack are key decision points in the battle, and commanders are encouraged to position themselves such that these decisions are not delegated. The depth attack should achieve maximum concentration of combat power against its key targets. If at all possible, security functions should be performed by other units in support of the depth attack group.

7-85. Depth attacks may employ one point of entry or multiple points, they may be staged and synchronized, and they may employ third-dimension mobility. The depth attack should be arranged to meet few obstacles and entrenched enemy forces in order to maximize the rapidity of movement into enemy rear areas. It may be supported by demonstrations, feints, assaults, or other penetrations designed to occupy the enemy's attention and divert its combat power. Figure 7-13 on page 7-28 depicts a depth attack.

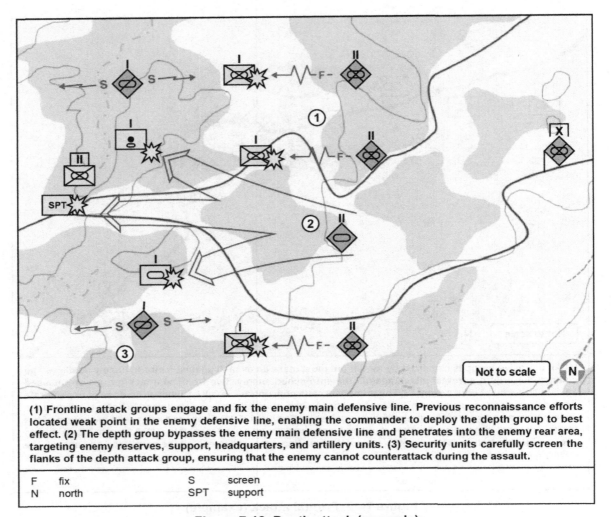

(1) Frontline attack groups engage and fix the enemy main defensive line. Previous reconnaissance efforts located weak point in the enemy defensive line, enabling the commander to deploy the depth group to best effect. (2) The depth group bypasses the enemy main defensive line and penetrates into the enemy rear area, targeting enemy reserves, support, headquarters, and artillery units. (3) Security units carefully screen the flanks of the depth attack group, ensuring that the enemy cannot counterattack during the assault.

| | | | |
|---|---|---|---|
| F | fix | S | screen |
| N | north | SPT | support |

**Figure 7-13. Depth attack (example)**

## Infiltration

7-86. Infiltration is an offensive tactic wherein a force closes with and conducts an undetected penetration of enemy defenses by using stealth, rapid movement, and deception. Infiltration tactics were the backbone of PLA tactics for much of the Chinese Civil War and Korean War, and they remain an important element of light infantry operations today. The primary benefit of infiltration tactics is that they offset enemy advantages in firepower. By closing distances and conducting penetrating maneuvers covertly, enemy firepower cannot be brought to bear prior to close contact. This enables friendly units to surprise the enemy, creating shock and the perception of isolation. The main limitation of infiltration is that it is very difficult to do with larger formations or with any unit that is not light infantry—large formations and vehicles are far easier to detect, and thus have great difficulty infiltrating enemy positions. Infiltration tactics are also very time consuming and limited in both scope and range, due to the slow movement rate of dismounted infantry and the need to maintain secrecy and concealment.

7-87. The PLAA today employs infiltration tactics alongside penetrations or other assaults to isolate and annihilate enemy units. For example, a light infantry depth attack group may infiltrate one flank of an enemy formation while an armor-heavy depth attack group conducts an aggressive penetration of another flank.

7-88. Successfully executing an infiltration attack requires considerable skill and initiative by small-unit leaders. These leaders are charged with identifying enemy weak points, then choosing their own avenues of approach and methods of attack. By its nature, an infiltration attack requires less concentration of forces than other forms of offensive action, but it requires greater skill in the advance and unfold phases of the operation, as detection by the enemy essentially changes the action from an infiltration to a frontal assault. Artillery and other enablers are also largely unavailable to an infiltration due to the need for secrecy and concealment, though information warfare and deception operations may provide vital support by fooling the enemy about the strength or direction of the attack or by directing enemy reconnaissance focuses elsewhere. Figure 7-14 depicts an infiltration.

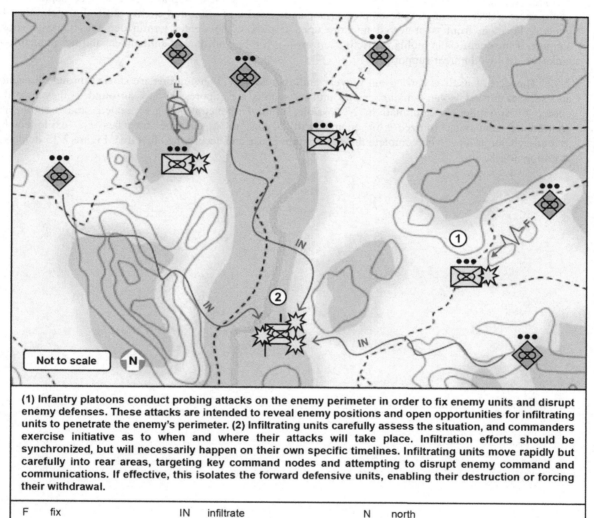

(1) Infantry platoons conduct probing attacks on the enemy perimeter in order to fix enemy units and disrupt enemy defenses. These attacks are intended to reveal enemy positions and open opportunities for infiltrating units to penetrate the enemy's perimeter. (2) Infiltrating units carefully assess the situation, and commanders exercise initiative as to when and where their attacks will take place. Infiltration efforts should be synchronized, but will necessarily happen on their own specific timelines. Infiltrating units move rapidly but carefully into rear areas, targeting key command nodes and attempting to disrupt enemy command and communications. If effective, this isolates the forward defensive units, enabling their destruction or forcing their withdrawal.

| F | fix | IN | infiltrate | N | north |
|---|-----|----|-----------|----|-------|

**Figure 7-14. Infiltration (example)**

## Storming Attack

7-89. A storming attack is a thoroughly prepared, high-intensity assault against an entrenched, isolated opponent. Storming occurs after penetrations defeat the enemy's main defensive line, most enemy units have been forced to retreat or been annihilated, and the PLAA unit must complete the annihilation phase by reducing the remaining enemy strongpoints. Storming attacks place great emphasis on obstacle defeat, firepower and maneuver integration, and aggressive action at the point of contact.

7-90. Storming attacks are preceded by a meticulous and integrated effort to clear enemy obstacles through a combination of firepower and engineering capabilities. These breaches provide the points of entry for the final assault. At the same time, careful reconnaissance of the enemy's position informs the commander of the enemy's strength and disposition. Firepower and information attacks against enemy positions are conducted throughout the operation, attempting to reduce enemy strength and cohesion, reduce morale, and suppress enemy systems. The enemy should be surrounded if possible, and it should be isolated not only physically, but also electronically and psychologically.

7-91. The main decision the commander must make in the assault phase is the deployment of armor. Three methods for armor deployment are prescribed: in front, in close support, and in rear support. The commander's decision on how to employ armor is dictated by terrain and enemy antitank capabilities. In general, tanks in front is preferred in more open terrain, while tighter terrain demands tanks in close support. If the enemy is in highly restrictive terrain—or if it has substantial antitank capabilities available—tanks are deployed in rear support.

7-92. The assault itself is carried out bravely and aggressively. Penetrations are carried through breaches and seek to further isolate smaller enemy units. As units are suppressed and surrounded, they face the choice of surrender or annihilation. If possible, multiple penetrations are created, and assaults are coordinated and synchronized. As enemy will begins to break, the breaches are exploited by second-echelon troops, who complete the annihilation phase and secure the objective. Figure 7-15 depicts a storming attack.

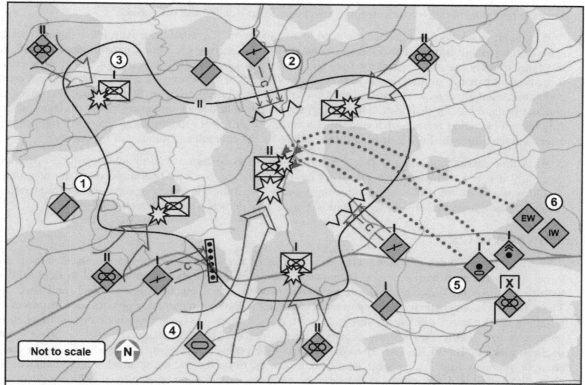

(1) Reconnaissance units carefully assess the surrounded enemy's position in order to discover potential weak points and establish the enemy's disposition. Once the attack commences, these units transition to screening the assaulting force. (2) Engineer units rapidly clear obstacles and minefields that lay along key avenues of approach. These efforts are carefully integrated with reconnaissance in order to develop the best possible course of action for the assault. (3) Frontline attack groups assault isolated and vulnerable forward enemy units. These attacks gradually increase in intensity as the enemy command is destroyed and isolation increases. (4) The depth attack group conducts a penetration into the enemy's center, destroying the enemy's command and completing the isolation of remaining enemy units. (5) The firepower group conducts mass fire assaults on the enemy's main body in order to disrupt enemy operations and reduce enemy morale. (6) EW and IW groups conduct electronic and information attacks on the enemy formation in order to amplify the effects of isolation and prevent communications with possible relief forces.

| C | clear | IW | information warfare |
|----|-------------------|----|---------------------|
| EW | electronic warfare | N | north |

**Figure 7-15. Storming attack (example)**

## PURSUIT ATTACK

7-93. Pursuit attack is an offensive tactic wherein a friendly formation maintains constant contact with an enemy formation in retrograde. This most often occurs when an enemy force is defeated but not surrounded, and it is thus able to retreat. Pursuit entails maintaining contact with the retreating unit and conducting quick attacks as required or where possible, with the objective of preventing effective enemy rear security operations—thus preventing the enemy unit from consolidating or reforming a strong defensive position.

7-94. Aggressively pursuing a defeated enemy formation is a basic component of mechanized warfare. The PLAA historically employed both mechanized and armored forces and fast-moving infantry forces in this role; on a modern battlefield, a mechanized force is likely required. Following an initial penetration, envelopment, or isolation, the enemy force is dislodged from its primary defensive position and begins retrograde operations to attempt to regain the initiative. Aggressive pursuit of the enemy allows friendly forces to retain the initiative while keeping the enemy force isolated and off balance.

7-95. Pursuit attack does not necessarily need to take the form of ground maneuver. Other capabilities can achieve the pursuit effects of maintaining contact with the enemy, preventing enemy consolidation, and retaining friendly initiative. Artillery fire, air attack, and countermobility activities can disrupt enemy formations or canalize or direct enemy movements. Network or electromagnetic attack can suppress enemy communications or disseminate disinformation to influence enemy actions. SOF or guerrilla forces can conduct raids, feints, or demonstrations to degrade enemy cohesion or influence enemy decision making. Any pursuit operation is best conducted using a variety of different capabilities to create a combined arms effect and maximize the impact of friendly operations on the strength and mindset of the enemy.

7-96. A unit conducting pursuit operations is vulnerable to counterattacks or other spoiling operations, especially if it is a maneuver unit strung out over a wide area. Security operations or supporting attacks should work to ensure that a unit conducting a pursuit operation is secure against enemy counterattack, reconnaissance, and countermobility activities. Figure 7-16 depicts a pursuit attack.

## FIREPOWER ATTACK

7-97. A firepower attack is a tactic wherein a number of firepower capabilities are massed on a specific target to achieve an offensive effect. Firepower capabilities include, but are not limited to field artillery, ballistic or cruise missiles, air attack, antiair attack, network attack, electromagnetic attack, and direct fire. The defining characteristic of the firepower attack is that it does not require any physical occupation of key terrain or close contact with enemy forces. Instead, massed firepower is used to isolate or annihilate enemy units, canalize enemy movements, penetrate enemy defenses, reduce enemy morale, or affect enemy decision making. The PLAA uses firepower attacks in a variety of ways, including to reduce cohesion or strength of enemy forces, to support assaults or advances, to fix the enemy, to attack enemy flanks or rear areas, or to spoil or defeat enemy counterattacks. Firepower attacks can be employed either alone or in an integrated and synchronized manner with other actions.

7-98. The PLAA recognizes at least seven different forms of firepower attack: advance fire, fire support, information and firepower assault, depth firepower, counter-counterattack firepower, annihilation firepower, and air defense firepower. These forms of firepower attack are described in paragraphs 7-99 through 7-107.

### Advance Fire

7-99. Advance fire is fire that targets enemy positions or other key assets prior to follow-on operations. In Western operations, it is typically referred to as preparatory fire. Advance fire targets enemy reconnaissance platforms, command and communication nodes, security units, and artillery units. The purpose of advance fire is to neutralize, destroy, or degrade key enemy capabilities in support of future operations.

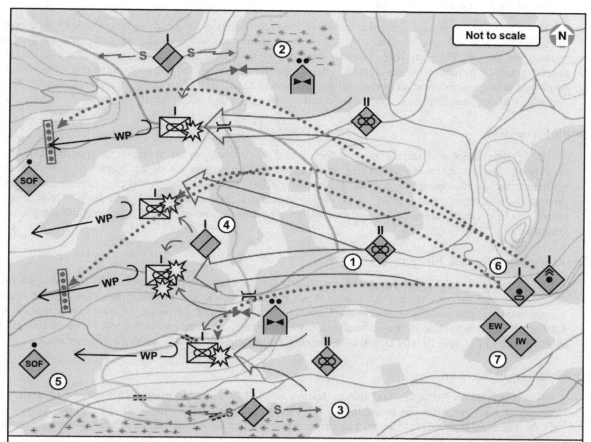

(1) An offensive group conducts a pursuit attack against a retreating enemy. CA-BNs maintain continuous contact with retreating enemy formations, not allowing then to conduct rearguard or delaying actions, and not allowing them to establish a new entrenched defensive position. (2) Attack helicopters are well-suited to support pursuit attacks; they continually harass and disrupt enemy units on the move. (3) Reconnaissance elements screen the flanks of the offensive group, ensuring that the enemy cannot mount a surprise attack on the extended friendly line. (4) Other reconnaissance elements maintain contact with the retreating enemy forces. (5) SOF teams in deep areas sabotage roads and bridges, disrupt enemy reinforcement efforts, and conduct deep reconnaissance. (6) The firepower group employs howitzers to attack retreating enemy units and rocket artillery to deploy hasty minefields across probably routes of retreat. (7) The EW group ensures that enemy forces cannot contact possible reinforcements, maintaining their isolation. The IW group concentrates on maximizing the propaganda value of the victory, transmitting media of destroyed and retreating enemy units.

| | | | | | | |
|---|---|---|---|---|---|---|
| CA-BN | combined arms battalion | N | north | SOF | special operations forces |
| EW | electronic warfare | S | screen | WP | withdrawal |
| IW | information warfare | | | | |

**Figure 7-16. Pursuit attack (example)**

## Fire Support (Advance or Assault)

7-100.    Fire support consists of those fires that are integrated closely with the advance phase or initiate phase of an operation. During the advance phase, the key mission for firepower is to ensure that enemy reconnaissance platforms are suppressed or destroyed—thus preserving the secrecy of the attack—along with enemy artillery units that may target the vulnerable advancing force during its movement operation. In support of assaults, firepower targets key enemy positions and units, creating weak points or breaches in the enemy's defensive line that enable effective penetration. Key targets of this mission type include enemy armor and antitank units, obstacles, fortified positions, and exposed units.

### Information and Firepower Assault

7-101. The information and firepower assault is also referred to as the blitz—a specialized tactic wherein physical fires and information attack are carefully synchronized to maximize the combined arms effect against the enemy. Physical fires destroy, degrade, or neutralize key targets through their destructive power, while information attack targets enemy morale, cohesion, and communications. The blitz is employed at the key point in an overall offensive operation, targeting the enemy's center of gravity. When and where to employ the blitz is a key consideration when planning offensive operations.

### Depth Firepower

7-102. Depth firepower targets enemy forces and key assets in deep areas, supporting follow-on operations that are exploiting successful initial assaults. Key targets for depth firepower include enemy artillery reserves attempting to spoil the exploitation effort, mobile enemy reserves conducting retrograde operations, enemy command posts, and enemy reconnaissance platforms. Depth firepower's primary mission is to create or enhance the isolation effect on enemy forces and enable the annihilation phase to commence.

### Counter-counterattack Firepower

7-103. Counter-counterattack firepower specifically targets enemy counterattacks. This includes not only targeting ground forces, but also air power through antiair attack and information attack against enemy communications and command structures. The counter-counterattack firepower mission is reserved for the point in time when the enemy's counterattack has been clearly detected and identified. It can be thought of as a miniature blitz, specifically targeting the enemy's commitment of reserves to the counterattack.

### Annihilation Firepower

7-104. Annihilation firepower attempts to destroy isolated or encircled troops. This may be in support of storming attacks, or it may be to annihilate the enemy independently. Annihilation firepower targets not only the enemy's isolated formations, but also any possible routes for retreat, any enemy forces attempting to relieve the besieged unit, and any obstacles that might inhibit a storming attack.

### Air Defense Firepower

7-105. Air defense firepower uses ground-based defensive counterair systems—surface-to-air missile systems, self-propelled antiaircraft guns, and EW systems—in an aggressive, forward-positioned manner. There are two possible missions for air defense firepower: restrict the enemy's use of a specific volume of airspace through deterrence, or destroy or attrit enemy aircraft. Though it employs systems that Western armies view as defensive, the PLAA characterizes air defense operations as fundamentally offensive, either on their own, or in support of other offensive actions.

7-106. In order to restrict the enemy's use of airspace, commanders should deploy their air defense systems aggressively, fully integrated with higher echelon air defenses and aircraft. When attempting to destroy or attrit enemy aircraft, commanders should use raid-like tactics, carefully concealing their air defense systems in order to surprise and overwhelm enemy aircraft. Missions can be performed separately or concurrently; a deterrence mission may be used to canalize aircraft toward a well-positioned air defense raid, for instance.

7-107. Air defense firepower is deployed in two distinct ways: to defend static assets, such as assembly areas, headquarters, or supply bases; or to defend units on the move. Air defense firepower is concentrated to protect the most important assets and oriented towards the enemy's most likely avenue of approach. Commanders employ a mix of air defense capabilities to achieve a combined arms effect. Air defense systems use several interlocking tiers of systems to offset friendly vulnerabilities.

### AMBUSH

7-108. An ambush is an operation in which a unit deploys to a location in advance, then strikes the enemy from a concealed position in a surprise attack. The PLAA historically emphasized ambush tactics as a highly effective way to offset enemy advantages in firepower or protection, and it used ambushes extensively during the Second Sino-Japanese War and the Chinese Civil War. The ambush is still a popular tactic, but the PLAA advises that the informationized battlefield makes conducting an effective ambush more challenging. Figure 7-17 on page 7-36 depicts an ambush operation.

7-109. The PLAA recognizes three distinct types of ambush: the waiting ambush, the decoy ambush, and the forced ambush. The waiting ambush is employed when an enemy travels along a predictable route, and friendly forces can deploy and lie in wait in a concealed position. A decoy ambush is similar, but instead of making use of a known route, it employs a decoy that tricks the enemy into moving into the ambush zone. A forced ambush is somewhat different: it employs direct actions such as feints, demonstrations, or countermobility activities to force the enemy into the ambush zone.

7-110. Choosing the ambush site is the most important decision when planning an ambush. Terrain should canalize the enemy formation while helping to conceal the friendly unit. If possible, enemy forces should be carefully reconnoitered to ensure friendly forces will enjoy local combat superiority. Engineers should set up obstacles to maneuver the enemy into the ambush zone and complicate its retreat. Artillery is deployed in concealed positions and registered across the ambush zone.

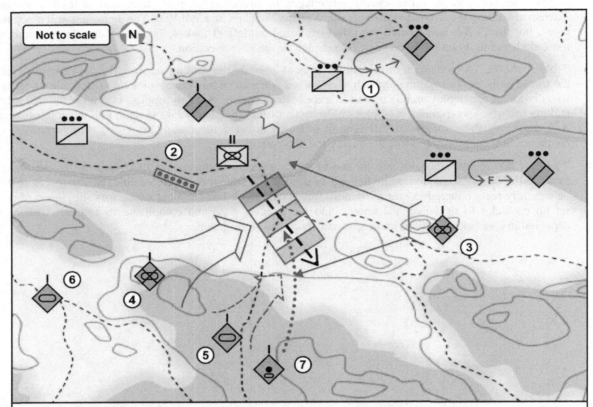

(1) A combined arms battalion sets a waiting ambush for an enemy battalion. Reconnaissance units carefully engage enemy scouts with feint attacks, drawing them away from the ambush zone and ensuring the main ambush force remains undetected. (2) Mines and obstacles laid along the enemy's potential routes are sued to canalize enemy movement into the ambush zone. (3) The concealed group occupies a well-protected support-by-fire position overlooking the ambush zone. This group initiates the ambush when the enemy enters the ambush zone. (4) The flank group occupies an adjacent position and engages the enemy's vulnerable flank once the enemy is heavily engaged by the concealed group. (5) The pursuit group waits just beyond the ambush zone and prepares to assault and pursue the enemy as it retreats from the ambush zone. (6) The interception group waits in a concealed position and prepares to aid in pursuit or intercept any reinforcements being sent to aid the ambushed unit. (7) The firepower group delivers preregistered artillery fire into the ambush zone.

| F | feint | N | north |
| --- | --- | --- | --- |

**Figure 7-17. Ambush (example)**

7-111. Maneuver units on ambush missions are divided into four primary groups: concealed, flank, pursuit, and interception. The <u>concealed</u> group initiates contact, engages, and fixes enemy forces from at least one flank. The <u>flank</u> unit assaults an alternate flank once the enemy force is fixed by the concealed group. The pursuit group is a reserve that engages and maintains contact with the enemy force as it retreats from the ambush zone. The <u>interception</u> group is deployed along potential reinforcement avenues of approach in order to prevent support or reinforcement to the ambushed enemy formation.

7-112. Concealment and deception are key enablers of the ambush. Reconnaissance and surveillance on the informationized battlefield make moving to ambush positions in secret a difficult task. Commanders must carefully move forces into their ambush positions, and units must remain highly disciplined in order to avoid detection. Once the ambush is complete and the enemy unit annihilated, secure retrograde operations should be conducted quickly.

## RAID

7-113. A raid is an offensive operation designed to strike and surprise an unsuspecting enemy. The primary defining characteristic of the raid is the employment of a hit-and-run approach. The enemy is struck and injured, then, in most cases, the attacking force breaks contact and rapidly retreats. Though raids may evolve into full-scale assaults in certain circumstances—encountering a weaker-than-expected enemy, for instance—the typical approach to a raid is a fast-moving, intentionally abbreviated attack.

7-114. Raids are aggressive actions that have clearly defined parameters. They may seek to harass the enemy, to probe it, or to destroy isolated enemy units. A raid requires skillful use of concealment and deception, rapid movement, and sudden, aggressive action within the commander's parameters. Subordinate units must be aware of the commander's intent and must not over-engage the enemy unless a clear opportunity presents itself.

7-115. The PLAA recognizes five different types of raids: rapid, long-range, sabotage, sneak, and harassing. These types of raids are described in paragraphs 7-116 through 7-120. Reconnaissance raids are discussed in more detail in chapter 6.

### Rapid Raid

7-116. The rapid raid is analogous to the Western idea of a meeting engagement. A unit unexpectedly comes upon an enemy, transitions to an attack posture, then conducts an aggressive attack that is pressed until the commander deems it complete. Rapid raids are specifically prescribed as spoiling actions against counterattacking forces or newly arrived airborne or amphibious forces.

### Long-range Raid

7-117. The long-range raid is not necessarily defined by the distance the unit travels to conduct the raid. Rather, it refers to how deep behind the enemy's main defensive line the raid takes place. Long-range raids are conducted by light forces or SOF who infiltrate the enemy's territory to strike at predetermined targets. This may be a separate action, or it may be integrated as part of a larger operation.

### Sabotage Raid

7-118. A sabotage raid is one in which a specific high-value asset is targeted for destruction. The unit infiltrates enemy territory with the intent of locating and destroying one or more high-value assets through covert, aggressive action. Potential targets include command nodes, key intersections or roadways, lines of communications, ammunition stockpiles, or high-value weapons systems such as combat aircraft or surface- to-air missile systems.

### Sneak Raid

7-119. The sneak raid is a small-scale, short-distance raid designed primarily to make use of local conditions such as nighttime or bad weather to conceal the raiding force. Sneak raids may attempt to seize enemy prisoners, probe enemy defenses, or reconnoiter. The sneak raid does not attempt to strongly engage the enemy, and the sneak raiders retrograde quickly back to concealed positions. Figure 7-18 on page 7-38 depicts a sneak raid against an enemy position.

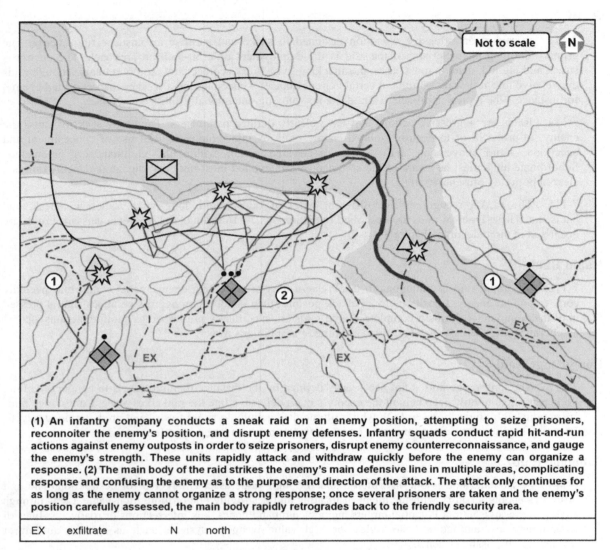

(1) An infantry company conducts a sneak raid on an enemy position, attempting to seize prisoners, reconnoiter the enemy's position, and disrupt enemy defenses. Infantry squads conduct rapid hit-and-run actions against enemy outposts in order to seize prisoners, disrupt enemy counterreconnaissance, and gauge the enemy's strength. These units rapidly attack and withdraw quickly before the enemy can organize a response. (2) The main body of the raid strikes the enemy's main defensive line in multiple areas, complicating response and confusing the enemy as to the purpose and direction of the attack. The attack only continues for as long as the enemy cannot organize a strong response; once several prisoners are taken and the enemy's position carefully assessed, the main body rapidly retrogrades back to the friendly security area.

| EX | exfiltrate | N | north |
|----|-----------|---|-------|

**Figure 7-18. Sneak raid (example)**

### Harassing Raid

7-120.    The harassing raid is one intended to annoy, disrupt, or confuse the enemy. Harassing raids are intended to provoke an enemy response, wear out enemy soldiers, or spoil enemy maneuvers. Most importantly, they are used to disperse the enemy, forcing it to defend a wider geographic area and reducing its presence at potential key places on the battlefield.

# Chapter 8

# Defensive Actions

A well-planned and -coordinated defense is a critical component of every combat action. The defense is carried out in order to attrit enemy forces, retain key positions or terrain, buy the commander time and decision space, seize the initiative from the enemy, and transition to offensive operations. The People's Liberation Army Army (PLAA) believes that defense is an inherently stronger form of war, as the defender enjoys advantages of terrain and time unavailable to the attacker. The informationized battlefield, however, has reduced many of these traditional advantages. While it is not possible to destroy an opponent through defensive actions alone, a tenacious and well-executed defense enables decisive follow-on offensive operations.

## OVERVIEW OF PLAA DEFENSIVE OPERATIONS

8-1.   The PLAA takes the position that defensive actions are ultimately to preserve one or more assets, such as friendly forces, key terrain, or the initiative. In addition, defensive operations can play a key role in a wider operational or strategic sense by attritting the enemy's strength, forcing it to commit greater forces in an attempt to achieve an objective, and reducing or restricting the options available to enemy commanders. Though the PLAA considers the offense to be the decisive form of warfare, centuries of invasion and occupation have led to considerable emphasis on the importance of defensive operations: the People's Liberation Army's (PLA's) most sacred mission is defending Chinese territory from outside aggression.

## THE INFORMATIONIZED BATTLEFIELD AND DEFENSIVE OPERATIONS

8-2.   The PLAA still maintains that defense is a fundamentally stronger form of warfare than offense, but it acknowledges that many elements of the informationized battlefield have changed the traditional dynamics between attack and defense. Most—though not all—of these changes benefit the attacker, making defensive operations more difficult than they have been historically. The PLAA has identified four major trends on the informationized battlefield that influence defensive operations: increasing arduousness, fewer traditional advantages, more dynamic, and increasing importance of offensive actions.

### INCREASING ARDUOUSNESS

8-3.   Multiple factors have combined to make combat for the defender more difficult than in the past. While ground commanders once only had to concern themselves with enemy land forces, enemies can now strike simultaneously or consecutively from multiple domains. This requires, in turn, a comprehensive multi-domain defense that can effectively blunt or check enemy actions, even when coming from unexpected directions. Commanders must also defend their forces not only from physical attack, but also from information, electromagnetic, and psychological attack: a capable opponent will target enemy troops' morale and cohesion through a variety of channels. Finally, the depth and variety of enemy firepower systems have increased substantially in recent years. Long-range artillery and missile strikes, air strikes, attack helicopter operations, direct action by special operations forces (SOF), and electromagnetic and network attack capabilities allow an enemy to target the full depth of friendly formations and defended areas. There is no safe space on the informationized battlefield.

## FEWER TRADITIONAL ADVANTAGES

8-4.   Historically, a defender enjoyed several fundamental advantages that translated to nearly every combat action: greater ability to conceal one's forces and deceive the enemy; better use of terrain to cover and harden one's position; improved communications and coordination due to interior lines and known terrain; and more time to make the best use of the deployment area and battlefield depth. Conditions on the informationized battlefield have eroded all of these traditional advantages. Because of advanced wide-area multispectrum reconnaissance, intelligence, and surveillance capabilities, concealing one's own forces and deceiving the enemy about deployments and dispositions is far more difficult. In the past, one only needed to win the ground reconnaissance and counterreconnaissance battle, but the informationized battlefield leverages air, space, and cyber intelligence collection in addition to more advanced ground-based capabilities. Precision munitions and long-range strike capabilities have eroded several basic defensive advantages offered by terrain, making it far more difficult to effectively harden defensive positions. The enemy can also effectively target the command and communication systems of a defensive position in ways never before available. Electronic warfare methods, such as communications jamming and meaconing, and network attack are capable of destroying, disabling, degrading, or manipulating the command and communication backbone of a defensive position. Finally, the defender's traditional advantage of depth has been eroded or eliminated by capabilities that enable the enemy to strike into deep areas. Depth areas that were once generally considered secure and untouchable can now be targeted with both lethal and nonlethal attack methods, even when many miles from forward areas.

## MORE DYNAMIC

8-5.   As modern offensive operations demand rapid thrusts that target weak points from unexpected directions, the traditional approach of defending strong points with a relatively static approach is now obsolete. Static formations, even in a strong defensive position, will quickly be bypassed, isolated, and then annihilated by a competent attacker. Instead, defense must be considered a dynamic action characterized by rapid movement and decisive concentration of combat power at key times and places. Information superiority coupled with mobility enables the defender to rapidly detect and appropriately respond to enemy offensive actions, regardless of where they might occur. This approach to defense is in keeping with the PLAA's new emphasis on decentralization. A mobile, rapidly reacting defensive strategy requires lower-echelon commanders able to react appropriately to unexpected events without input from a higher echelon. While the hasty defense was once considered a last-ditch option, it is today the cornerstone of a PLAA defensive operation versus a capable opponent.

## INCREASING IMPORTANCE OF OFFENSIVE ACTIONS

8-6.   While seeming somewhat paradoxical, offensive actions within a broader defensive operation are considered to be of heightened importance. Due to the conditions of the informationized battlefield, offensive actions have increased in their lethality and unpredictability; this trend extends to counterattacks. A well- timed and -executed counterattack can disrupt, defeat, or spoil an enemy's offensive action just as effectively as a well-planned defense, and the counterattack enjoys all the advantages of other offensive actions: enhanced firepower, enhanced reconnaissance, multi-domain options, and the elements of surprise and initiative. The PLAA approach to defense-by-offense prescribes the use of depth attacks and aggressive maneuver to put the counterattack group in the best possible position, concentrating combat power against enemy weak points and enabling the isolation and destruction of enemy offensive groups.

# PRINCIPLES OF THE DEFENSE

8-7.   The PLAA has identified four key principles that should guide commanders' planning and operations when conducting a defensive operation on the informationized battlefield: depth, consolidation, integration, and flexibility. When applied skillfully, these principles enable the commander conducting a defensive action to effectively blunt the opponent's attack and then rapidly transition to decisive offensive operations.

## DEPTH

8-8.    While defense-in-depth is an ancient concept, its modern application requires a somewhat different approach than the traditional ground-only concept. Defense-in-depth now requires PLAA commanders to integrate defensive measures throughout all domains while placing a strong emphasis on the counterattack. It is impossible to defend all areas of the modern battlefield from all potential avenues of attack, so commanders must effectively prioritize key areas for defense and make best use of available space— trading space for time is now more important than ever before. Commanders should encourage enemies to ingress through unimportant or noncritical areas, extending their lines of communications (LOCs) and exposing their flanks. Decisive counterattacks should target these formations at key times and places, seeking to disrupt or destroy enemy formations as they move through deep areas. These efforts are simultaneously aided by raids, firepower operations, and information warfare operations that target enemy reinforcement movement into the combat area. A properly executed defensive operation makes use of depth to achieve isolation of enemy forces while effectively integrating all available assets into a cooperative defensive effort throughout the combat area.

## CONSOLIDATION

8-9.    The defensive principle of consolidation mirrors the offensive principle of concentration: it prescribes massing combat power at key times and places to make best use of available forces. Effective consolidation is underpinned by the thoughtful deployment of friendly forces to key points around the defensive zone before wider defensive operations commence. This deployment, in turn, requires the commander and staff to conduct extensive analysis of the enemy's probable approach, then prescribe the primary defensive line and axis, <u>key defense points</u> (KDPs), and key targets for the counterattack. Defensive operations must take advantage of favorable terrain or information conditions as combat multipliers, as counterattack groups are expected to be outnumbered by the forces they engage. The main defensive combat capability should take the form of a powerful mobile force able to conduct independent operations anywhere in the defensive zone. The commitment of this force typically represents the key point in a defensive battle. In addition, powerful mobile and armored forces should seek to conceal their movements as much as possible to deceive the enemy about the time and place of their commitment, and they should remain as dispersed as possible to protect against enemy firepower and to further deceive the enemy.

## INTEGRATION

8-10.    A defensive operation must be carefully integrated throughout the defensive zone. This includes not only careful planning and unification of purpose, but also preservation of combat power, effective use of terrain and firepower as means of resistance, and use of small-unit actions across the defensive zone to harass and distract the enemy. Protection of friendly forces is a key element of integration. This involves not only traditional hardening of defensive positions, but winning the information battle as well—keeping friendly forces from being affected by enemy psychological attack and propaganda. Physical resistance should be enhanced by the use of obstacles and other countermobility capabilities, which are used to slow, stop, or canalize enemy formations in accordance with the overall scheme of defense. The appropriate use of firepower and terrain can allow vastly overpowered units the ability to resist enemy attacks for long periods, forcing the enemy to commit reserves and overextend its offensive to achieve its attack goals. Antitank and air defense weaponry must be fully integrated into the defense in order to offset enemy advantages in armor, firepower, and air power. Coordinated firepower assaults mixing direct fire and artillery can attrit or disrupt enemy formations at stand-off distances. Depth operations target enemy LOCs and reinforcements, attempting to isolate powerful enemy offensive groups as they penetrate into the defensive zone.

## FLEXIBILITY

8-11. The defensive principle of flexibility seeks to achieve a state of <u>localized initiative</u> where defensive forces, despite fighting on the defensive, successfully maneuver, consolidate, and achieve localized combat power at key times and places. It requires commanders to take prudent risk in their defensive plans, to use stratagems effectively in order to gain the advantage of surprise, and to attack and defend resolutely as the situation develops. A flexible defense is able to absorb enemy offensive actions by trading time and space, then rapidly adjusting to enemy tactics and strategy to formulate a response. A flexible defense is, in many ways, a series of small offensive actions, seeking to target enemy weak points with friendly strength and seizing localized initiative at key times and places. The sort of flexibility that the PLAA seeks requires superior lower-echelon leadership, with lower echelons capable of independent actions and lower echelon commanders able to make good decisions even in the absence of complete information.

# PLANNING THE DEFENSE

8-12. PLAA doctrine prescribes a large number of steps for planning and executing a defensive operation, which can be consolidated into seven major phases:

- Build the command system.
- Organize reconnaissance.
- Organize the defensive group and deploy.
- Spoil the enemy's preparations.
- Resist the enemy's assaults.
- Counterattack.
- Consolidate or withdraw.

Each phase is part of every defensive operation to some degree, but the phases are not necessarily sequential. Because they are largely in response to enemy actions, they may be conducted in varied order or concurrently. This planning includes organizing the defensive battlefield and building the defensive groups that comprise the operational system.

## BUILD THE COMMAND SYSTEM

8-13. As with offensive operations, the PLAA prefers to carefully plan and organize major defensive operations. Meticulous planning enables greater redundancy and depth in the defense, improved overall security measures, a more effective deception plan, and a greater chance of a decisive counterattack. The centerpiece of the defensive plan is the deployment of the command posts. A defensive operation makes use of at least two, and ideally four, command posts. The base command post is the commander's primary location, and it will typically be located to best coordinate between frontier defense and depth defense units, oriented toward the enemy's anticipated main axis of attack. The commander may also establish an advance command post in a more forward position. The rear command post is led by the deputy commander, and it is deployed in a well-defended location. Its primary role is to organize logistics and rear area defense. If possible, a reserve command post is established along a possible route of egress or in a well-defended rear location, ready to take over for the base or advance command post should either of them come under threat. Additional information on command post operations can be found in chapter 4.

8-14. The defensive battle makes careful use of terrain and geographic control. There are two primary features that a commander must identify to make best use of the terrain. First, the main direction of defense informs the combat group as to the orientation of the enemy's anticipated primary effort. Commanders placing the main direction of defense should take into account higher echelon and adjacent-unit missions, specified defensive tasks, the unique characteristics and tendencies of the opponent, and terrain throughout the defensive zone. Second, the commander must establish KDPs. These are the specific features within the defensive zone that are of greatest importance to the integrity of the overall defense. Commanders should keep the numbers of KDPs small, choosing only those points absolutely critical to successfully prosecuting the defensive battle. The defensive plan should be centered on maintaining these KDPs, which assists commanders in prioritizing their available resources. KDPs should also be phased, allowing for them to change during the course of the battle. Finally, these points are not limited to key terrain. They may also include a critical command or network node, a key unit or leader, or even a piece of information on a computer network.

8-15. The defensive zone is further subdivided into two or more secondary zones, each with a specific set of objectives and different set of tactics. While the PLAA used to be highly prescriptive about the physical sizes of these zones, it has gradually moved to a more flexible approach. The various zones should account for terrain, friendly and enemy capabilities, and higher echelon missions. They should also enable careful integration of various units and capabilities. Figure 8-1 depicts a combined arms brigade's (CA-BDE's) genericized defensive zone

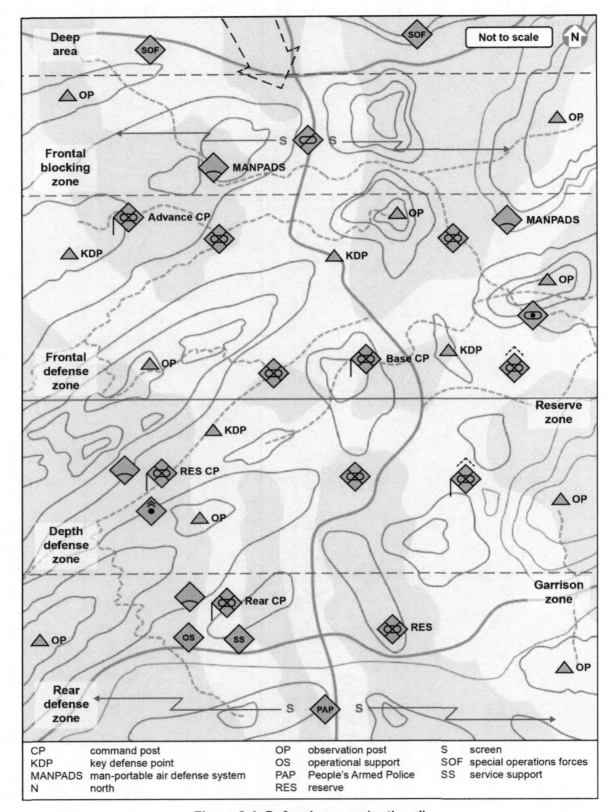

**Figure 8-1. Defensive zone (notional)**

| CP | command post | OP | observation post | S | screen |
|---|---|---|---|---|---|
| KDP | key defense point | OS | operational support | SOF | special operations forces |
| MANPADS | man-portable air defense system | PAP | People's Armed Police | SS | service support |
| N | north | RES | reserve | | |

### Deep Area

8-16. Deep areas are those areas that are not targetable by a defensive group's organic weapons systems. They may still be occupied by elements of the defensive group conducting reconnaissance, counterreconnaissance, and screening missions, but these elements operate independently. SOF elements also occupy deep areas, and both air and missile support may be employed in support of deep-area operations. The primary purpose of deep area operations is to disrupt and slow the enemy advance while providing the defensive group commander with critical intelligence about the enemy's strength, disposition, and possible objectives.

### Frontal Blocking Zone

8-17. The frontal blocking zone is the forward-most area of the defensive zone. It is analogous to the Western security zone, and it is designed in much the same way. The frontal blocking zone is occupied by units performing screen or cover missions, and it serves as the primary early warning, disruption, reconnaissance, and counterreconnaissance zone for the main body. The frontal blocking zone should orient toward the enemy's primary axis of advance—particularly its primary reconnaissance axis. This zone should be positioned sufficiently in front of the main body to give the commander time and decision space, usually between 3 kilometers (km) and 5 km for a CA-BDE. The frontal blocking zone is likely occupied primarily by the cover group.

### Frontier Defense Zone

8-18. The frontier defense zone is typically the combat group's primary defensive area. It should contain most KDPs and the preponderance of combat power. Units should seek to conduct a strong defense within the frontier defense zone, attempting to force the enemy to commit most of its combat power and leave its forces vulnerable to counterattack. The frontier defense zone is primarily occupied by one or more frontier defense groups, and possibly a depth group. Operations in this zone are centered largely on active resistance to enemy assaults. Fortifications and entrenchments should be as extensive as time and resources allow, and units should be prepared to conduct stubborn and brave resistance, possibly in the face of significant odds. The base command post is typically located in this zone, while the advance command post will be either forward in the zone or rearward in the frontal blocking zone.

### Depth Defense Zone

8-19. The depth defense zone extends behind the frontier defense zone, and it serves as the deep area of the defensive zone. Depth defense groups—the heart of the counterattack force—are deployed here, to react quickly and appropriately to the enemy's assaults in the forward defensive areas. In addition, combat reserve groups may be stationed in this zone to reinforce or assist forward units. A single depth defense zone may serve as the deep area for multiple frontier defense zones. Operations in the depth defense zone should ensure protection of the depth defense group from air or artillery assault, ensure mobility, and ensure the concealment and secrecy of counterattack operations. The reserve command post, if established, will likely be located in this zone.

### Rear Defense Zone

8-20. The rear defense zone is the deepest area of the defensive zone, and it houses logistics, equipment support, and other rear area units and capabilities, including the rear command post. This zone is occupied primarily by rear area security units—possibly People's Armed Police (PAP) units in addition to PLAA security forces—and it may also house depth or combat reserve groups. The rear defense zone ensures protection against enemy deep artillery, air strikes, and enemy SOF or guerrilla actions in rear areas, and it ensures mobility to enable retrograde or reinforcement movements through the zone.

### ORGANIZE RECONNAISSANCE

8-21. The reconnaissance effort is a critical enabler of a defensive operation. Reconnaissance must be comprehensive and well planned, extending through all domains and the depth of the battlefield. These efforts concentrate on three primary intelligence objectives: disposition and intent of enemy forces; terrain;

and conditions of the battlefield—civilians, weather, the electromagnetic environment, and so on. Reconnaissance commences immediately upon receipt of warning orders: commanders rapidly build and deploy the reconnaissance and intelligence group to facilitate rapid, accurate, and continuous scouting for the main body. Like all other aspects of a defensive battle, reconnaissance must be conducted in depth. Deep and forward reconnaissance units surveil and disrupt enemy forces in deep areas. As the situation develops, additional reconnaissance forces, having been held in reserve, are deployed to key areas of the battlefield to enhance situational understanding.

8-22. Reconnaissance groups are often charged with disrupting enemy operations, typically as part of a screen or cover force. This requires a mixture of stubborn resistance and prudent tactical offensive actions, usually in the form of raids coupled with entrenched units. These activities are intended to force the enemy to deploy earlier than it would like, buying the commander decision space and revealing the enemy's plan of action. At the same time, counterreconnaissance actions attempt to spoil the enemy's efforts to ascertain friendly force dispositions, leaving it ignorant as to the strong and weak points of the defense. Reconnaissance screens may attempt to canalize or steer enemy forces toward areas of strength.

8-23. Reconnaissance information should be carefully analyzed and processed to give the commander the clearest possible picture of the battlefield. The commander must identify the most important elements of information, and the reconnaissance and intelligence group must strive to answer these questions as clearly and accurately as possible. Intelligence filtering enables the commander to judge friendly and enemy situations; make best use of terrain, weather, and the conditions of the electromagnetic spectrum; then distill the ground truth from the "fog of war." In contrast to its historical approach of top-down intelligence, the post-reform PLAA places greatest value on the lower-echelon commander's situational understanding, above that which is handed down by upper-level leadership. Lower-level commanders are encouraged to act decisively based on their own judgment, though they must consider the higher-echelon mission and intent. Additional information on reconnaissance and security operations and the PLAA's intelligence process can be found in chapter 6.

## ORGANIZE THE DEFENSIVE GROUP AND DEPLOY

8-24. The combat power that comprises the defense is organized into a <u>defensive group</u>, the combat group assigned to the defensive mission. This operational system is built in the same way as any other tactical operational system, using the CA-BDE as the primary force provider and the combined arms battalion (CA-BN) as the primary building block. Capabilities are organized into one of four primary groups, each with a distinct mission: the frontier defense group, the depth defense group, the combat reserve group, and the cover group. A combat group may contain some or all of these, and it may contain more than one of each. Additional groups that may be employed include a firepower strike or artillery group, an air defense group, an electronic and network warfare group, and a combat support group. A specific description of each of these groups can be found in chapter 4. In addition, the commander may designate teams for specific missions, such as antitank, mobile artillery, obstacle construction or reduction, or rear defense.

8-25. Once the defensive group is assembled, groups deploy to the defensive zone. This action should be undertaken quickly, allowing time for adjustments and amendments to orders after the groups reach their initial positions. The cover group typically has the mission of scouting, occupying, and securing the defensive area, enabling the other groups to move into place quickly and safely. Any enemy reconnaissance or scouting elements in the defensive area must be aggressively attacked and driven off, ensuring secrecy of movement and disposition of friendly forces. These operations are likely performed in tandem with ongoing friendly reconnaissance efforts.

8-26. Secrecy of movement is a critical consideration as follow-on friendly forces occupy their assigned defensive positions. Commanders must always assume that the enemy is watching, and they should employ concealment and deception as much as possible to confuse or deceive the enemy. Commanders must also be aware of the threats posed by artillery bombardments, airborne or ground-based raids, and information operations efforts targeting the combat group throughout this phase.

8-27. Engineering construction begins immediately as the groups enter the defensive position. Engineer priorities are determined primarily by terrain, enemy capabilities, and time available. A hasty defense may only allow simple structures, such as improvised obstacles and firing positions. With more time available, a more elaborate defense may allow for extensive tunnels, entrenchments, field structures, permanent shelters, and fallback positions. The PLAA emphasizes the use of irregular units such as local militia and local populations for engineering construction efforts. All engineering efforts focus heavily on KDPs. Figure 8-2 depicts reconnaissance and deployment activities occurring in the defensive zone.

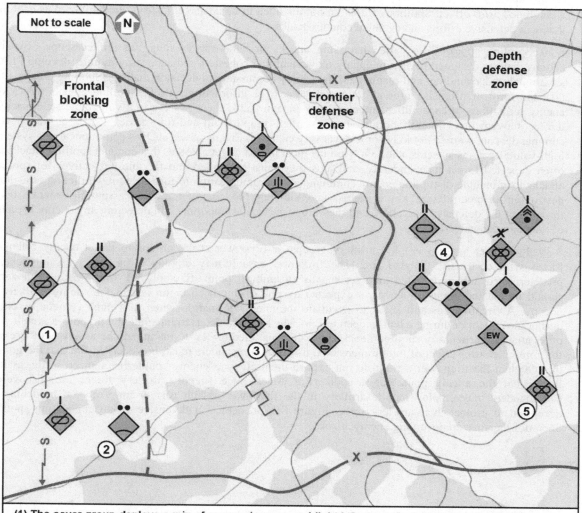

(1) The cover group deploys a mix of reconnaissance and light infantry units into the frontal blocking zone. These units conduct reconnaissance and counterreconnaissance missions while screening the main body. (2) MANPADS sections establish ambush zones along potential aerial avenues of approach. (3) The frontier defense zone is occupied by the frontier defense group. The main line of defense consists of two key defense points, each defended by a mechanized CA-BN. SPG batteries and SPAAG sections are in direct support. (4) The depth defense group consists of two armored CA-BNs supported by a heavy towed battery, a rocket battery, and a SHORAD platoon. This group is charged with conducting the decisive counterattack against the enemy's main effort. (5) The reserve group consists of a mechanized CA-BN. It awaits orders to block enemy advances or to support the depth group's counterattack.

| CA-BN | combined arms battalion | SHORAD | short-range air defense | S | screen |
|---|---|---|---|---|---|
| MANPADS | man-portable air defense system | SPG | self-propelled gun | SPAAG | self-propelled anti aircraft |
| N | north | EW | electronic warfare | | |

**Figure 8-2. Reconnaissance and deployment in the defensive zone (example)**

## SPOIL THE ENEMY'S PREPARATIONS

8-28. Once the reconnaissance efforts determine the enemy's presence and disposition, the defensive group commences efforts to spoil the enemy's offensive plan. In most cases, this phase involves two primary efforts: spoiling attacks and firepower assaults. Ideally, these efforts are conducted concurrently to achieve a combined-arms effect. Spoiling efforts are not intended to be decisive; instead, they are seen as enabling the main defensive efforts and eventual counterattacks.

8-29. Spoiling attacks, also called harassing attacks or harassing assaults, are limited-scope offensive actions that are designed to disrupt enemy movement and cohesion, reduce enemy morale, develop PLAA situational understanding, and manipulate the enemy commander's decision making. Spoiling attacks may take the form of offensive ground action, air attack, or information attack. Spoiling attacks target key enemy formations or capabilities, attempting to engage them when they are in a vulnerable state, such as during movement. Ground spoiling attacks often employ hit-and-run tactics—such as raids—and commanders are warned not to become decisively engaged. Enemy assets such as communications systems, high-value weapons systems, and LOCs are the most prized targets. Forces that conduct spoiling attacks are often expected to act independently, and thus they must be skilled and flexible. Objectives for spoiling attacks may include disrupting enemy command and communication to forward units, inflicting casualties, destroying or neutralizing key systems, and slowing the enemy's advance. Spoiling attacks occur throughout the defensive operation, and they are of particular importance in defeating any enemy breaches of the main defensive line.

8-30. Firepower assaults employ artillery groups or firepower strike groups to target enemy offensive forces with destructive massed fire. The firepower assault may employ a firepower assault zone, a predetermined and pretargeted area serving as an ambush point. If employed, firepower assault zones should be located along the enemy's expected avenue of approach or on terrain the enemy is likely to occupy. A firepower assault seeks to decimate the enemy formation when it is most vulnerable, either during movement or during a halt in open terrain. The best practice is to integrate the firepower assault with other attacks, particularly spoiling attacks by reconnaissance forces, to maximize the accuracy of fire and the combined-arms effect of multidimensional threats. Commanders must, however, ensure that the groups that conduct the firepower assault do not expose themselves to enemy counterfire unnecessarily, as it is likely that the enemy plans for a counterfire attack once friendly artillery forces are discovered. Commanders may employ mobile artillery in the firepower assault role or may rely on hardening or deception to protect their artillery force. Figure 8-3 on depicts a defensive group conducting spoiling operations against an upcoming enemy attack.

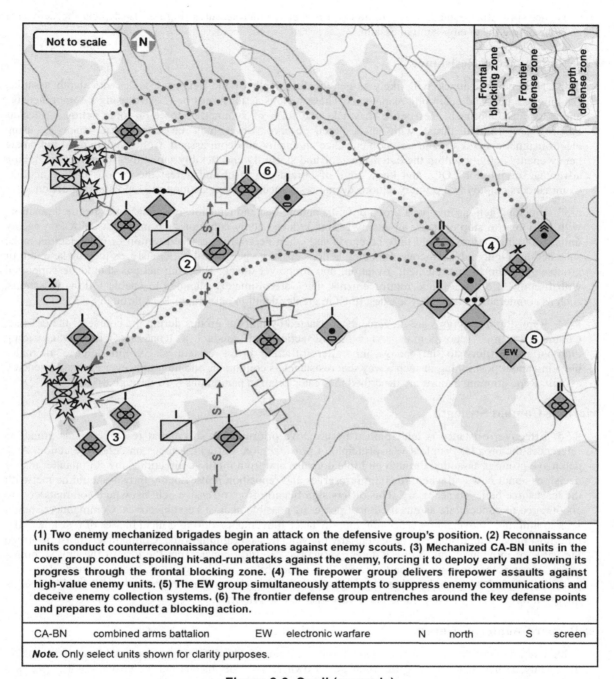

(1) Two enemy mechanized brigades begin an attack on the defensive group's position. (2) Reconnaissance units conduct counterreconnaissance operations against enemy scouts. (3) Mechanized CA-BN units in the cover group conduct spoiling hit-and-run attacks against the enemy, forcing it to deploy early and slowing its progress through the frontal blocking zone. (4) The firepower group delivers firepower assaults against high-value enemy units. (5) The EW group simultaneously attempts to suppress enemy communications and deceive enemy collection systems. (6) The frontier defense group entrenches around the key defense points and prepares to conduct a blocking action.

| CA-BN | combined arms battalion | EW | electronic warfare | N | north | S | screen |

*Note.* Only select units shown for clarity purposes.

**Figure 8-3. Spoil (example)**

## RESIST THE ENEMY'S ASSAULTS

8-31. The main defensive effort occurs when the enemy commences its main assault through the defensive zone. The enemy's attack consists of a mixture of maneuver, firepower, and information attack, coordinated and synchronized to overwhelm, destroy, or force the withdrawal of friendly units. The focus of the resistance phase is to blunt the enemy's attack; sap its combat power, cohesion, and morale; and put it in a vulnerable position. This allows the defender to seize the initiative through aggressive counterattacks and then transition to the offense. The PLAA anticipates fighting throughout the depth of the defensive zone, and it offers three guidelines to inform commanders during this critical phase: wage simultaneous resistance, prioritize combat strength, and use proper countermeasures. Figure 8-4 depicts the defensive group resisting the enemy's initial assault.

### Wage Simultaneous Resistance

8-32. The enemy's attack will likely target the full depth of the defensive zone with depth assaults, flanking assaults, and encircling maneuvers. These various actions must be resisted using a comprehensive and integrated approach. Resistance may take the form of symmetric, force-on-force actions—such as meeting an armored thrust with an armored reserve—or it may be asymmetric, such as using electromagnetic attack to confuse and neutralize an enemy air depth assault. Commanders must anticipate heavy enemy activity within the defensive zone and skillfully and flexibly move their own forces—taking advantage of interior LOCs and knowledge of terrain—to meet and defeat the various enemy actions. Commanders employ two primary actions during resistance: blocking actions and repositioning actions.

8-33. Blocking actions involve a group or unit standing fast and tenaciously defending a specified position, with the intent to stop or delay an enemy action, reduce enemy cohesion, and inflict casualties on enemy units. Blocking actions should only be conducted when necessary, or when terrain or other factors make such actions highly advantageous. Blocking actions are to be performed judiciously, as they make a unit or group vulnerable to encirclement, isolation, and firepower attack. Blocking actions should be supported with firepower, obstacles, and reinforcements. They are ultimately intended to enable follow-on actions, such as counterattacks, and they are described in greater detail in paragraphs 8-73 through 8-75.

8-34. Repositioning actions are movements conducted by units or groups during the course of the defense. Commanders mix repositioning and blocking actions to conduct a tenacious withdrawal, ceding unimportant territory to the enemy only after inflicting heavy casualties. Skillful defense alternates blocking and repositioning in such a way that resistance is continuous and no unit is ever exposed to enemy assault. Repositioning actions are described in greater detail in paragraphs 8-76 through 8-78.

### Prioritize Combat Strength

8-35. Effective resistance is underpinned by effective prioritization of combat resources. This guideline references the broader People's War principle of prioritization, but with a focus on defensive action. As a defensive group is assaulted through multiple domains and from multiple directions, the commander makes decisions about how and where to commit reserves and reposition units. Enemy thrusts should be met with the least force believed necessary. This differs significantly from offensive actions, where commanders are encouraged to concentrate as much combat power as possible against key objectives. Commanders should seek to minimize movement—especially long-distance movement—during critical phases of resistance. If a unit must move long distances, it is essentially unavailable as a fighting force. Commanders must prioritize the value of moving a unit during a defensive action, carefully considering the loss of the unit's combat power during the movement period and the vulnerability of the unit to enemy action. Effective prioritization is largely underwritten by high-quality, comprehensive reconnaissance and intelligence, as quality information tells the commander how and where to prioritize.

### Use Proper Countermeasures

8-36. Even the best defensive action is inherently reactionary. The defensive group must assess and react to the enemy, rather than dictating the fight on its own terms. This means that defensive combat places a higher premium on employing the right tactical means to counter enemy capabilities. Commanders must have a clear picture of the enemy capabilities they are facing and build operational systems to effectively counter their anticipated opponent. So too must commanders rapidly and effectively react to the enemy's actions as the situation unfolds, deploying friendly combat power to counteract enemy penetrations and assaults.

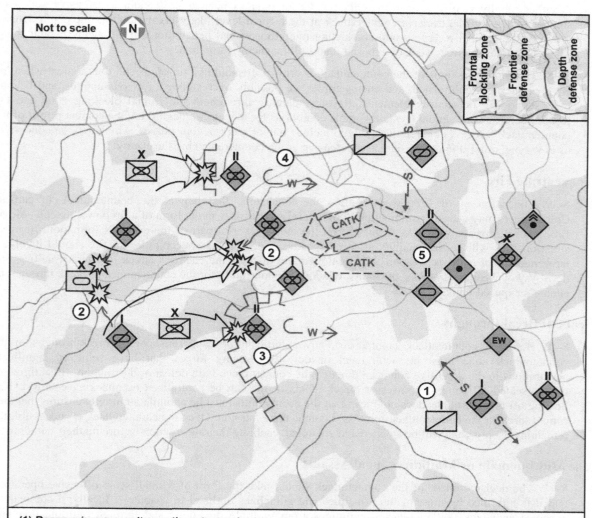

(1) Reconnaissance units continue to conduct counterreconnaissance operations against advancing enemy scouts to prevent detection of the disposition and axis of the main counterattack. (2) The cover group commences spoiling attacks and raids against the enemy's main effort, causing disruption and forcing early deployment. These attacks are continuous and intended to canalize the enemy's main effort, directing it to the point in the frontier defense zone where the main counterattack is planned. (3) The frontier defense group conducts strong resistance against enemy attacks on key defensive positions. The enemy attempts to fix the two CA-BNs and destroy them in detail; entrenchments and firepower blunt the attack. (4) The CA-BNs conduct a retrograde action, falling back slowly and blocking enemy penetrations into rear areas. (5) The depth group begins to move toward the counterattack position. By this time, the enemy's position and axis of advance are well known, and the commander carefully chooses the time and place for the counterattack.

| CA-BN | combined arms battalion | EW | electronic warfare | S | screen |
| CATK | counterattack | N | north | W | withdraw |

*Note.* Only select units shown for clarity purposes.

**Figure 8-4. Resist (example)**

## COUNTERATTACK

8-37. The counterattack, also called the mobile assault, is typically the culmination of a defensive operation. It requires the defensive group to mass combat power and conduct an aggressive, decisive attack against one or more enemy units, with the intent of disintegrating the enemy attack and forcing the enemy to either rapidly retreat or face annihilation. Counterattacks happen throughout a defensive action on smaller scales, but the decisive counterattack at the tactical level likely involves either the depth defense group or the combat reserve group conducting multidimensional penetrations of an enemy assault force. These penetrations target weak or exposed flanks of the enemy formation.

8-38. The timing and axis of the counterattack is critical. It should be timed around the moment when the enemy attack has culminated, and enemy forces are possibly overextended, low on supplies, and beyond their supporting or security elements. Resolute defense throughout the frontal blocking zone and the frontier defense zone should create favorable conditions for the attack, magnifying the combat power of the counterattack force. A counterattack should include one or more of four phases, which may be concurrent or in varying order. Figure 8-5 depicts a defensive group counterattacking the enemy.

### Concentrate Fire

8-39. Concentration of firepower is often the touchstone that begins the counterattack in earnest. Commanders may use any combination of indirect fire—possibly in the form of a firepower assault—direct fire, and information attack, concentrated on the most powerful part of the enemy's formation. In many situations this will be the enemy's armored spearhead, and so the concentration of fire should involve a large number of antitank weapons. Concentrated fire disrupts the enemy's attack, creates casualties, and destroys key weapons systems, opening a window of opportunity for the counterattack to break through the enemy's exposed flank.

### Seal Off Breakthroughs

8-40. Before a counterattack commences, any significant or dangerous enemy breakthroughs in the counterattack zone must be neutralized or defeated. Enemy forces that have achieved significant penetrations in the defensive zone can threaten the counterattack with defeat or destruction. Breakthroughs can be sealed off using firepower or direct attack, or the can be neutralized by attacking enemy LOCs. Sealing off breakthroughs does not mean that the enemy units must be annihilated; it only ensures that they cannot threaten the counterattack force as it conducts its mission. Once the counterattack is successful, the remaining enemy units will be isolated and encircled, enabling their destruction or forcing their withdrawal.

### Use Multi-domain or Multidirectional Assault

8-41. The decisive phase of the counterattack should take the form of a small-scale offensive operation, and it is informed by most of the principles and guidelines outlined in chapter 7. Ideally, a commander executes an attack that involves penetrations from multiple directions and through multiple domains, preventing the enemy from massing combat power and confusing it about the disposition and axis of the main effort. Direct assaults on the enemy's front are the preferred method of attack, as this fixes the enemy and immediately targets its most valuable units. Pincer movements are considered a secondary approach, and while they may be effective at targeting the enemy's exposed flank, they require longer movement and more time, which exposes the counterattack force to enemy artillery or direct attack. Firepower assaults can cut off or block access to the rear, while blocking units work to stop further penetrations. The assault targets the enemy's center of gravity, attempting to isolate enemy units throughout the defensive zone.

### Hold Key Defense Points

8-42. As the counterattack commences, commanders assess its effectiveness. If it is effective, they may order a continuation of the attack or consolidation of the gains. If it is ineffective, they may order the counterattacking force to assume a defensive posture and resist any further enemy advances. Regardless of effectiveness, commanders must immediately deploy the defensive group to retain KDPs, be they still held by friendly forces or recently retaken. Commanders should anticipate the enemy sending reinforcements or conducting supporting attacks promptly after it recognizes the threat of the counterattack. Having the defensive group deployed and in place to block these attacks ensures that the gains achieved by the counterattack are not lost.

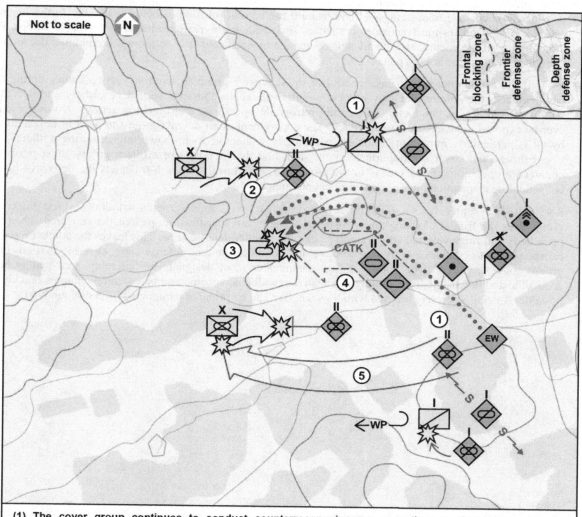

(1) The cover group continues to conduct counterreconnaissance operations against enemy scouts. Mechanized infantry elements of the cover group conduct attacks against enemy scouts, forcing their withdrawal and exposing the flanks of the enemy assault group. (2) The frontier defense group continues to conduct blocking actions against the enemy's supporting attacks, preventing penetrations and slowing the enemy's advance. (3) Firepower is concentrated on the enemy's main effort. Tube and rocket artillery plus electronic attack are massed on the target in order to disrupt movement, reduce cohesion, and cause casualties. (4) The depth group conducts its counterattack assault. Despite being outnumbered, the depth group achieves local superiority through mobility, deception, and the effective use of firepower. The depth group assaults the enemy's main body head on, attempting to destroy the enemy's momentum and cohesion. (5) The combat reserve group is deployed in an attack on the enemy's exposed flank in an attempt to isolate the enemy's main body and force the enemy's withdrawal.

| | | | | | | |
|---|---|---|---|---|---|---|
| CATK | counterattack | N | north | WP | withdrawal under pressure |
| EW | electronic warfare | S | screen | | |

*Note.* Only select units shown for clarity purposes.

**Figure 8-5. Counterattack (example)**

## CONSOLIDATE OR WITHDRAW

8-43. During or after the counterattack commanders face a critical decision on how to proceed with the defensive battle. If the counterattack succeeded in decimating, blunting, or neutralizing the enemy assault, commanders may consider ordering a follow-on attack into the enemy's depth. Alternatively, they may consider holding fast, reinforcing and entrenching the current position to better resist any further enemy assaults. Commanders must consider casualties sustained, the overall readiness of available troops, the vulnerability of the enemy, and the advantages gained by each course of action when making a decision to continue advancing or hold in place.

8-44. If the counterattack was unsuccessful or the defensive group suffered serious casualties during the counterattack, the commander may order a withdrawal. The withdrawal should be orderly and decisive, with available groups conducting alternating retrograde operations and blocking actions. No unit should ever be exposed to an enemy pursuit attack. The deployment of air defense groups to deter or defeat attack by air is particularly important. Units are particularly vulnerable to firepower attack during withdrawals. Concealment and cover are important for the unit, and the firepower or artillery groups must prioritize counterfire operations if any unit is in the process of withdrawal. Figure 8-6 depicts the defensive group consolidating its position after a successful counterattack.

8-45. The PLAA prescribes a rough order of precedence for a combat group's withdrawal. First, the cover group exits the frontal blocking zone. Next, all support and logistics groups exit the rear area. Then the main body and firepower and artillery groups exit the frontier defense zone. This leaves the combat reserve, or rear group, to conduct a rearguard or screening action in the rear defense zone, ensuring that all other groups move to safety. The combat group commander must designate an assembly area that is both accessible and defensible, then rapidly establish a new defensive zone and begin entrenching as quickly as possible. A detailed description of a withdrawal action can be found in paragraphs 8-88 through 8-91.

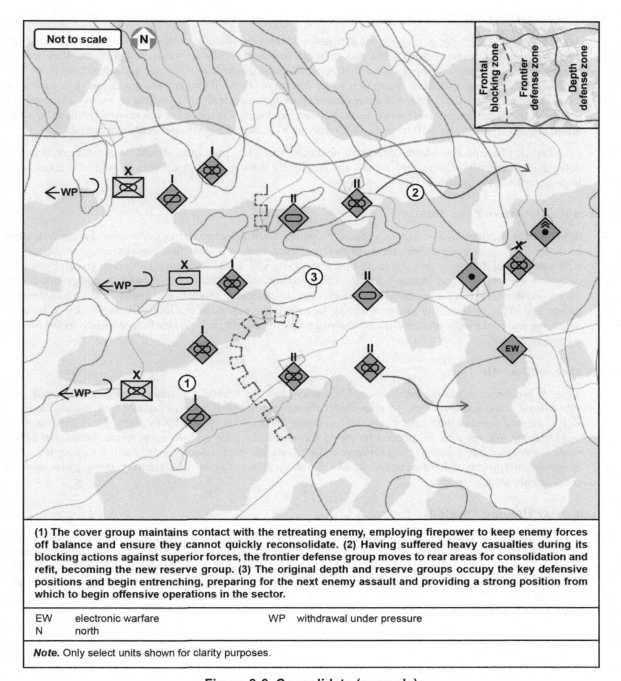

(1) The cover group maintains contact with the retreating enemy, employing firepower to keep enemy forces off balance and ensure they cannot quickly reconsolidate. (2) Having suffered heavy casualties during its blocking actions against superior forces, the frontier defense group moves to rear areas for consolidation and refit, becoming the new reserve group. (3) The original depth and reserve groups occupy the key defensive positions and begin entrenching, preparing for the next enemy assault and providing a strong position from which to begin offensive operations in the sector.

| | | | |
|---|---|---|---|
| EW | electronic warfare | WP | withdrawal under pressure |
| N | north | | |

*Note.* Only select units shown for clarity purposes.

**Figure 8-6. Consolidate (example)**

# TYPES OF DEFENSIVE OPERATIONS

8-46. The PLAA recognizes several types of defensive operations: positional, mobile, hasty, urban, diversionary, and specialized. These types of operations are described in paragraphs 8-47 through 8-69

## POSITIONAL DEFENSIVE OPERATIONS

8-47. <u>Positional defensive operations</u> are defensive operations that rely on a strongly entrenched position, defensive terrain advantages, obstacles, and stubborn resistance. They are static in nature, and they are not designed to enable the rapid relocation of troops. Positional operations are not considered decisive, and they should always be executed as part of a wider defensive action. Positional actions defend a key piece of terrain or other target, delay or block enemy forces moving through an area, deplete an enemy's strength by forcing it to assault entrenched positions, or buy time and decision space for offensive operations in other areas. Figures 8-7 through 8-10 on pages 8-19 through 8-22 depict a positional defensive operation.

8-48. When planning a positional defensive operation, three key principles should be considered: keep the defensive zone small, prepare sufficiently, and emphasize holding positions. These principles are described in paragraphs 8-49 through 8-51.

### Keep the Defensive Zone Small

8-49. Forces conducting a positional defensive operation are not expected to move with the same rapidity as those conducting more mobile forms of defense. This, in turn, means that concentrating combat power with a positional defensive group is more time consuming and difficult than with a mobile group. Positional defensive operations must ensure adequate concentration of combat power through a higher density of troops—rather than relying on mobility—which requires defensive zones to be smaller relative to defensive group size. This restricts the enemy's ability to maneuver, enhances the mutually supporting effects of defensive entrenchments, and enables defensive groups to operate with greater independence in the face of an enemy assault.

### Prepare Sufficiently

8-50. A positional defensive battle should be anticipated and carefully planned. Commanders must give subordinate units sufficient time to plan their defense and then build the necessary fortified positions on KDPs. Higher echelon defensive plans must account for the time necessary to prepare a defensive zone, and delaying actions may be needed to create the necessary time and decision space. Defensive group commanders should employ entrenchments, fortifications, and underground facilities, coupled with integrated firepower and information attack. Building these systems and creating these plans takes a considerable amount of time.

### Emphasize Holding Positions

8-51. A positional defensive operation should be built around holding strong pieces of terrain, even when outnumbered, under assault, or faced with intense firepower attack. Holding these pieces of terrain forces the enemy to commit and exhaust more and more troops in its attempt to break the defensive zone. With sufficient time and quality planning, a defensive group commander can create a comprehensive, in-depth plan that can hold strong positions, attrit the enemy, reduce the enemy's morale and cohesion, and buy other commanders in the area time and decision space.

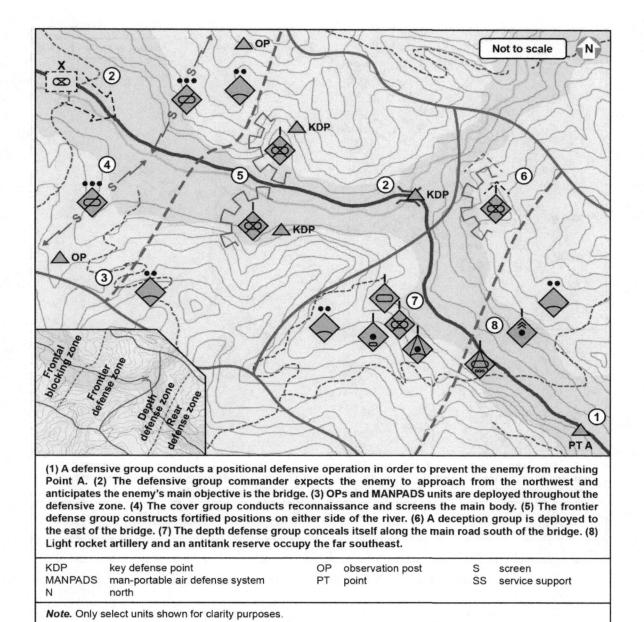

(1) A defensive group conducts a positional defensive operation in order to prevent the enemy from reaching Point A. (2) The defensive group commander expects the enemy to approach from the northwest and anticipates the enemy's main objective is the bridge. (3) OPs and MANPADS units are deployed throughout the defensive zone. (4) The cover group conducts reconnaissance and screens the main body. (5) The frontier defense group constructs fortified positions on either side of the river. (6) A deception group is deployed to the east of the bridge. (7) The depth defense group conceals itself along the main road south of the bridge. (8) Light rocket artillery and an antitank reserve occupy the far southeast.

| | | | | | |
|---|---|---|---|---|---|
| KDP | key defense point | OP | observation post | S | screen |
| MANPADS | man-portable air defense system | PT | point | SS | service support |
| N | north | | | | |

*Note.* Only select units shown for clarity purposes.

**Figure 8-7. Positional defense (example part 1 of 4)**

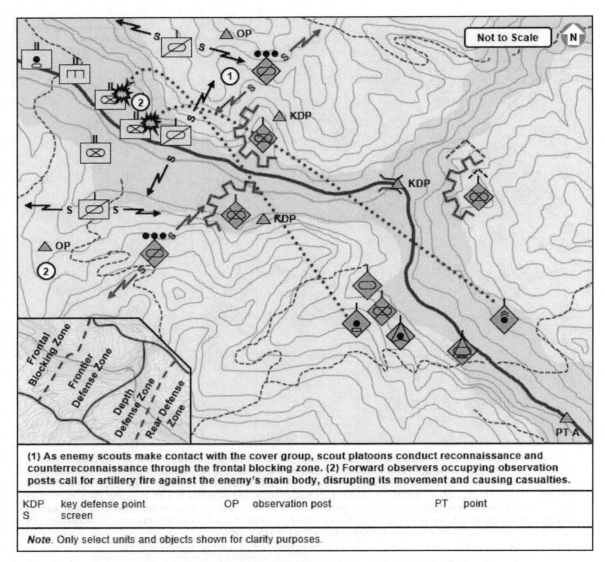

**Figure 8-8. Positional defense (example part 2 of 4)**

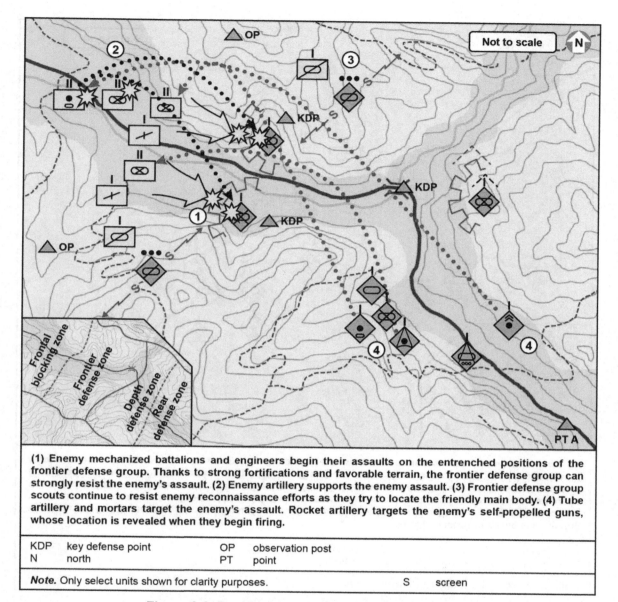

(1) Enemy mechanized battalions and engineers begin their assaults on the entrenched positions of the frontier defense group. Thanks to strong fortifications and favorable terrain, the frontier defense group can strongly resist the enemy's assault. (2) Enemy artillery supports the enemy assault. (3) Frontier defense group scouts continue to resist enemy reconnaissance efforts as they try to locate the friendly main body. (4) Tube artillery and mortars target the enemy's assault. Rocket artillery targets the enemy's self-propelled guns, whose location is revealed when they begin firing.

| | | | |
|---|---|---|---|
| KDP | key defense point | OP | observation post |
| N | north | PT | point |

| | | | |
|---|---|---|---|
| *Note.* Only select units shown for clarity purposes. | | S | screen |

**Figure 8-9. Positional defense (example part 3 of 4)**

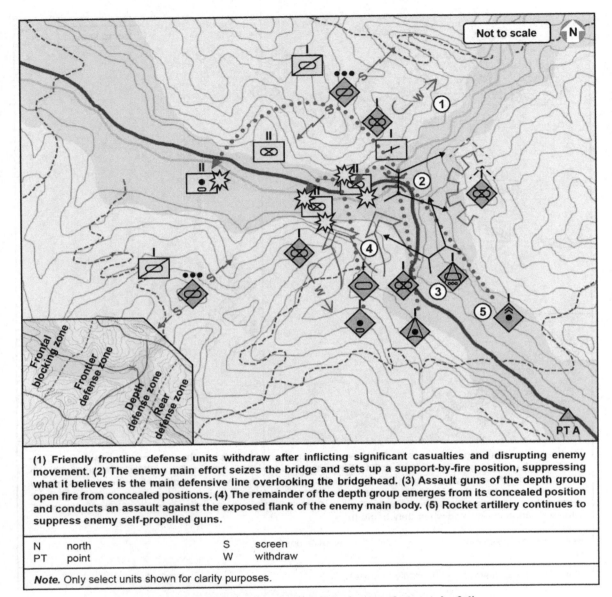

(1) Friendly frontline defense units withdraw after inflicting significant casualties and disrupting enemy movement. (2) The enemy main effort seizes the bridge and sets up a support-by-fire position, suppressing what it believes is the main defensive line overlooking the bridgehead. (3) Assault guns of the depth group open fire from concealed positions. (4) The remainder of the depth group emerges from its concealed position and conducts an assault against the exposed flank of the enemy main body. (5) Rocket artillery continues to suppress enemy self-propelled guns.

| N | north | S | screen |
|---|-------|---|--------|
| PT | point | W | withdraw |

*Note.* Only select units shown for clarity purposes.

**Figure 8-10. Positional defense (example part 4 of 4)**

## MOBILE DEFENSIVE OPERATIONS

8-52. Mobile defensive operations are defensive operations that mix blocking and restraining actions with counterattacks, seeking to achieve a decisive defeat of an opponent's assault. Mobile defensive operations employ armored or mechanized forces that are able to maneuver and concentrate combat power decisively at key points throughout the defensive zone. Firepower, information warfare, and engineer support are all carefully integrated to support the main counterattack effort. The mobile defensive operation seeks to halt the enemy assault, isolate attacking enemy units, then either annihilate them or force their withdrawal. Figures 8-11 through 8-14 on pages 8-24 through 8-27 depict a mobile defensive operation.

8-53. When planning a mobile defensive operation, five key principles should be considered: plan and execute quickly and decisively, assume a large defensive zone, focus on the offense, emphasize flexibility and mobility, and assume command and communication challenges. These principles are described in paragraphs 8-54 through 8-58.

## Plan and Execute Quickly and Decisively

8-54. Unlike positional defensive operations, mobile defensive operations are often opportunistic or hasty in nature, as they are fundamentally reactionary to enemy actions. This requires mobile defensive actions to be planned and executed on a much shorter timeline than positional defenses. Commanders must be flexible in their deployments and plan in advance as much as possible. They also must allow for changes to occur during the battle and allow the plan to adapt to the situation. Plans should, at a minimum, instruct lower-echelon units about key defensive positions, the main direction of defense, and how to employ key countermeasures to enemy actions.

## Assume a Large Defensive Zone

8-55. Mobile defensive operations should be conducted in a relatively wide geographic area. This forces the enemy to disperse its forces—making its intentions clearer earlier in the battle—and takes full advantage of the mobility of the defensive group. A larger relative area decreases the value of key defensive positions and makes defensive engineering more challenging and less effective. This, in turn, means that the bulk of the defense in a mobile operation falls to mobile units and their ability to concentrate combat power. The key competency of a mobile defense is the rapid transfer of combat strength, along with the ability to gracefully fall back or aggressively attack, as required.

## Focus on the Offense

8-56. The mobile defensive operation, though defensive at the operational level, is built around attack at the tactical level. An effective mobile defense emphasizes offensive operations, using the bulk of available combat power on attacks and counterattacks and seeking a decisive engagement with the enemy assault force. Other activities in the defensive zone should revolve around the decisive action by either manipulating enemy forces or providing direct support to the main counterattack. Mobile defensive operations must also employ a variety of combat actions, such as raids, ambushes, screens, firepower assaults, and information warfare to best support the counterattack effort.

## Emphasize Flexibility and Mobility

8-57. The success of a mobile defensive operation is contingent on the defensive group massing combat power at key times and places. Considering the large geographic area of the defensive zone and the assumption that the enemy possesses greater local combat power than the defensive group, a combination of flexibility and mobility are required to achieve decisive concentration. Commanders must be prepared to shift their center of gravity and conduct offensive actions across the flanks, depth, or rear areas of enemy formations, or anywhere else within their defensive zone. Defensive groups in mobile operations are often ad hoc or informal. Troops and units must be comfortable with the rapid forming and reforming of combat groups to fit the needs of the immediate situation.

## Assume Command and Communication Challenges

8-58. The combination of hasty action, a large defensive area, ad-hoc combat groups, and aggressive enemy actions make command and communication of a mobile defensive operation very difficult. Commanders must approach this issue with two different solutions: First, they must train their subordinates to act appropriately in the absence of orders or direct connection with their leaders. Second, they must maximize efforts to establish and sustain command and communication structure throughout the defensive area. Despite these efforts, commanders must assume that subordinate units will have to fight when isolated or out of contact, and they must provide their subordinates with the means to conduct operations without direct communication.

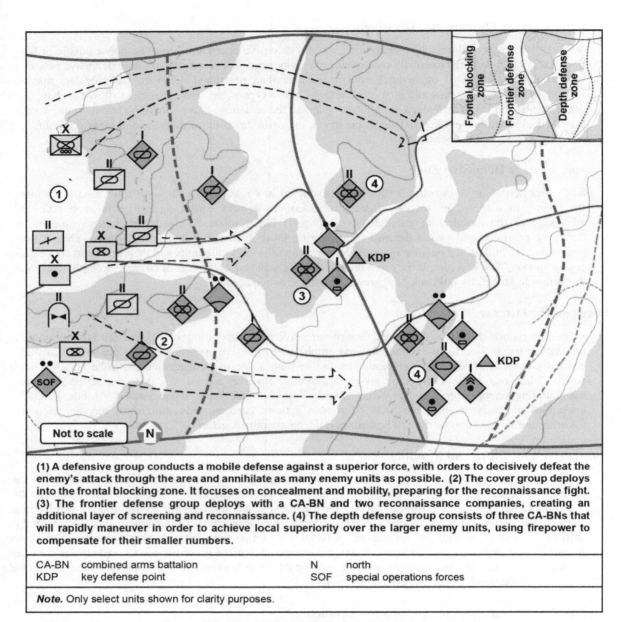

(1) A defensive group conducts a mobile defense against a superior force, with orders to decisively defeat the enemy's attack through the area and annihilate as many enemy units as possible. (2) The cover group deploys into the frontal blocking zone. It focuses on concealment and mobility, preparing for the reconnaissance fight. (3) The frontier defense group deploys with a CA-BN and two reconnaissance companies, creating an additional layer of screening and reconnaissance. (4) The depth defense group consists of three CA-BNs that will rapidly maneuver in order to achieve local superiority over the larger enemy units, using firepower to compensate for their smaller numbers.

| CA-BN | combined arms battalion | N | north |
| KDP | key defense point | SOF | special operations forces |

*Note.* Only select units shown for clarity purposes.

**Figure 8-11. Mobile defense (example part 1 of 4)**

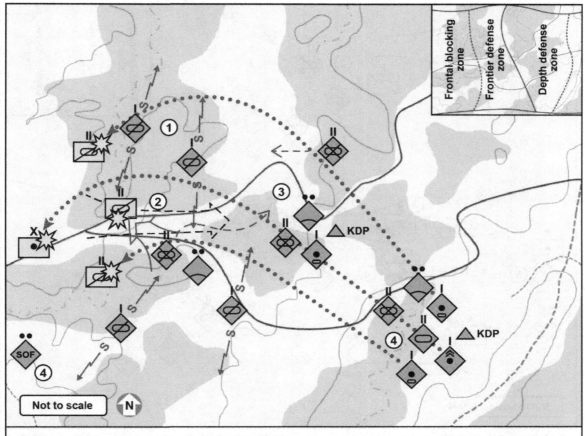

(1) The reconnaissance/counterreconnaissance fight commences as cover group units screen the main body and obstruct enemy scouting efforts. Scout units conduct a defense-in-depth while firepower attacks against enemy scout forces disrupt their movements. (2) The cover group's medium CA-BN conducts a surprise raid against the enemy's center, badly damaging an enemy scout unit. (3) The mechanized CA-BN then moves to join an impromptu defensive line with two other CA-BNs. (4) Aided by SOF forward observers, the heavy rocket battalion targets the enemy's artillery brigade with counterfire. The enemy cannot respond, as the 300-mm rockets significantly outrange its lighter rocket systems.

| CA-BN | combined arms battalion | | S | screen |
|---|---|---|---|---|
| N | north | | SOF | special operations forces |

*Note.* Only select units shown for clarity purposes.

Figure 8-12. Mobile defense (example part 2 of 4)

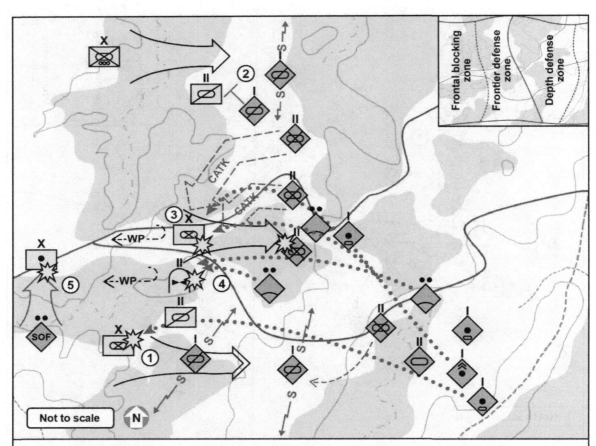

(1) The enemy's main effort commences to the south. Cover group units continue to screen against a vastly superior enemy force, attempting to slow and disrupt the main assault through stubborn resistance and firepower. (2) In the north, the enemy's flank penetration is met with a combination of screening and blocking actions, designed to divert the enemy northward and protect the central counterattack. (3) In the center, the enemy, deprived of effective scouting, assaults what it believes to be an inferior force. The defensive group commander has concentrated combat power in the center, and conducts a decisive counterattack against the flank of the enemy brigade. The counterattack, coupled with a firepower attack, compels an enemy withdrawal. This isolates the two remaining wings of the enemy force. (4) Medium-range air defenses force the enemy's aerial attack to lower altitudes, where an ambush consisting of MANPADS and anti-aircraft guns is waiting. The ambush inflicts severe damage on the enemy aviation unit, compelling its withdrawal and leaving the main effort without its air component. (5) SOF elements in enemy rear areas conduct a raid against enemy artillery, disrupting its efforts to conduct fire support and counterfire.

| CATK | counterattack | N | north | SOF | special operations forces |
|------|---------------|---|-------|-----|---------------------------|
| MANPADS | man-portable air defense system | S | screen | WP | withdrawal under pressure |

*Note.* Only select units shown for clarity purposes.

**Figure 8-13. Mobile defense (example part 3 of 4)**

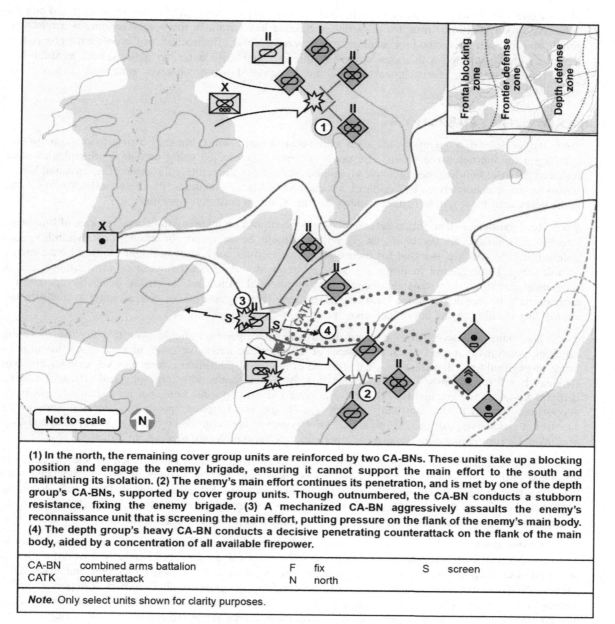

(1) In the north, the remaining cover group units are reinforced by two CA-BNs. These units take up a blocking position and engage the enemy brigade, ensuring it cannot support the main effort to the south and maintaining its isolation. (2) The enemy's main effort continues its penetration, and is met by one of the depth group's CA-BNs, supported by cover group units. Though outnumbered, the CA-BN conducts a stubborn resistance, fixing the enemy brigade. (3) A mechanized CA-BN aggressively assaults the enemy's reconnaissance unit that is screening the main effort, putting pressure on the flank of the enemy's main body. (4) The depth group's heavy CA-BN conducts a decisive penetrating counterattack on the flank of the main body, aided by a concentration of all available firepower.

| | | | | | |
|---|---|---|---|---|---|
| CA-BN | combined arms battalion | F | fix | S | screen |
| CATK | counterattack | N | north | | |

*Note.* Only select units shown for clarity purposes.

**Figure 8-14. Mobile defense (example part 4 of 4)**

## HASTY DEFENSIVE OPERATIONS

8-59. <u>Hasty defensive operations</u> are those defensive operations conducted under emergent circumstances and on a short timeline, usually in direct response to enemy actions. Hasty operations attempt to make best use of terrain and reconnaissance and security operations—coupled with surprise and deception—to defeat, neutralize, or blunt enemy advances. Hasty operations may take the form of abbreviated positional operations, mobile operations, or contain characteristics of both, depending on available forces, terrain, and enemy actions.

8-60. Hasty defensive operations are highly dependent on two key actions: extensive reconnaissance and security and the seizing and fortification of key terrain. The importance of reconnaissance is enhanced due to the shortened timeline available for planning and preparation, as the defensive group has less time to plan for enemy contingencies and entrench, while security efforts can delay the enemy assault and buy the defensive group additional time for preparation. Seizing key terrain in the defensive zone is a relatively simple and time-saving method for increasing the overall combat effectiveness of the defensive group. Key terrain should be selected with a view toward simple and aggressive defensive actions, such as ambushes. Once terrain is taken, entrenchments should be constructed using whatever time and resources are available.

## URBAN DEFENSIVE OPERATIONS

8-61. Urban defensive operations are seen as optimal in many ways, but they also provide some unique challenges for the defensive group. The urban defense is conducted using the same principles as other forms of defense: building the defensive group and establishing the primary direction of defense and KDPs. Urban terrain is characterized by buildings, streets, and subsurface facilities; planners must account for and make best use of these terrain features when planning an urban defensive operation.

8-62. Reconnaissance in an urban defensive zone concentrates on mapping the characteristics of buildings, streets, and subsurface developments. Buildings should be assessed for strength, vulnerability, and defensibility. Streets are assessed for their effects on mobility and how they might be used by an enemy assault force. Underground facilities are carefully scouted and assessed for use as fortifications and as movement corridors; they must also be assessed as potential vulnerabilities. If the defensive operation is anticipated to be lengthy, scouting must also assess long-term support assets, such as water availability and supply-line viability in and out of the defensive area.

8-63. The main defensive direction in an urban defensive operation should prioritize the use of strong blocking positions that can control wide areas of terrain in the defensive zone. In most cases this implies large, strong buildings that can be further fortified or hardened. Elevation is seen as a key enabler in urban defense. If a force controls the high ground, that force can command a large physical area while being difficult to detect and engage. Key defensive positions should be spread out, but they should be mutually supporting. Support points, consisting of supply, fallback, and firepower positions, are established to the rear of key defensive positions. Multiple defensive positions are supported by each support point, and they are linked by communications trenches or other concealed and protected methods of transit.

8-64. Urban defensive operations are more likely to involve complications, such as the presence of civilians, militia, or security forces. Commanders must plan for the various problems associated with civilian presence while integrating militia and security forces into the defensive plan. They must also make clear to subordinates the rules governing the presence and treatment of civilians in the defensive zone. Security forces may be leveraged to assist with the removal or protection of civilians.

8-65. Obstructions and obstacles are often more effective in urban areas due to the canalizing effects of streets and the lack of available alternative routes. The defensive group should make best use of obstacles to manipulate enemy movement, encouraging the enemy to overextend past its support and leave itself vulnerable to counterattack. Obstacles should be carefully integrated with firepower. Firepower that is optimized to engage an enemy attempting to breach an obstacle creates a powerful combined arms effect. It should also be used to dominate key avenues of approach. Firepower in urban terrain is often magnified by the close proximity of the combatants and the canalizing nature of the terrain, but it may be degraded by the wide availability of cover and protection offered by strong buildings and underground facilities. Figure 8-15 depicts an urban defensive operation.

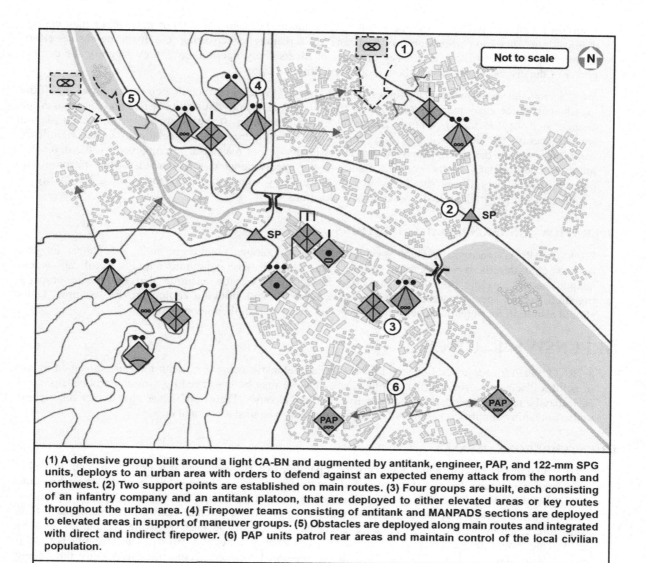

(1) A defensive group built around a light CA-BN and augmented by antitank, engineer, PAP, and 122-mm SPG units, deploys to an urban area with orders to defend against an expected enemy attack from the north and northwest. (2) Two support points are established on main routes. (3) Four groups are built, each consisting of an infantry company and an antitank platoon, that are deployed to either elevated areas or key routes throughout the urban area. (4) Firepower teams consisting of antitank and MANPADS sections are deployed to elevated areas in support of maneuver groups. (5) Obstacles are deployed along main routes and integrated with direct and indirect firepower. (6) PAP units patrol rear areas and maintain control of the local civilian population.

| CA-BN | combined arms battalion | MM | millimeter | SP | support point |
| M | meter | N | north | SPG | self-propelled gun |
| MANPADS | man-portable air defense system | PAP | People's Armed Police | | |

**Figure 8-15. Urban defensive operation (example)**

## DIVERSIONARY DEFENSIVE OPERATIONS

8-66. <u>Diversionary defensive operations</u> include the multitude of different temporary or hasty missions conducted both in and out of the defensive zone to surprise, confuse, frustrate, or degrade enemy forces. Missions under the diversionary umbrella include screens and covers, counterreconnaissance efforts, ambushes, blocking actions, and raids. Diversionary operations are defined by their purpose. They are intended to force the enemy to deploy, deplete enemy supplies, inflict casualties, slow the enemy advance, suppress enemy reconnaissance, and provide early warning for the main defensive group. They are typically conducted far in front of the main defensive zone. The deeper the penetration into the enemy's territory, the more effective these operations tend to be.

8-67. Depth is the key principle during diversionary defensive operations. Deploying in depth enhances the effects of frustration and confusion upon the enemy, as it requires enemy forces to continually fight through stubborn resistance. Diversionary operations may be along the main direction of defense, or they may be deployed in alternative directions as a means of screening or covering the main defensive effort. Countermobility and firepower are key enablers of diversionary operations, as they magnify the combat power of the diversionary force without requiring additional troops.

8-68. Commanders of diversionary forces probably have significant latitude in their decision making. If the diversionary force encounters a strong enemy force, it may decide to remain concealed and reconnoiter the enemy, or it may decide to carefully harass the enemy formation. If the diversionary force encounters a weaker enemy, the defending unit may aggressively assault the enemy, seeking to inflict significant casualties or spoil the enemy's movement. Diversionary forces may also be employed by the defensive group to counterattack on the enemy's flanks as the enemy penetrates forward defensive positions. In general, diversionary forces should avoid being decisively engaged, always maintaining the ability to rapidly and safely withdraw to deeper defensive positions.

## OPERATIONS UNDER SPECIAL CONDITIONS

8-69. The PLAA recognizes eight special defensive operations conditions: hills, mountains, jungles, extreme cold, deserts, rivers, swamps, and nighttime. Each of these conditions requires modifications and adaptations to baseline operations. Primary considerations include ensuring that troops are properly prepared for the special condition—with cold weather gear or bridging equipment, for example. Units should also train for potential special conditions that may occur in their combat area.

# DEFENSIVE TACTICS

8-70. The PLAA uses several different tactics during defensive operations, with their use informed by the principles discussed earlier in this chapter. Each of these may be employed by virtually any echelon, and each may be used independently, sequentially, or simultaneously. Figures in this section depict genericized units conducting simplified variants of the defensive tactics against a notional enemy.

## COUNTERRECONNAISSANCE

8-71. Counterreconnaissance as a tactic crosses operations of all types, and it is critical in each. Effective counterreconnaissance during defensive operations denies the enemy knowledge about the disposition and strength of friendly forces, the axis and timing of counterattacks, and the locations of key objectives. Counterreconnaissance may be active—attempting to aggressively defeat enemy reconnaissance efforts using direct action against reconnaissance elements—or it may be passive, using techniques such as camouflage, concealment, or decoys to fool or spoof enemy collection efforts. The best practice is to employ both active and passive measures in an integrated manner. Active measures deny the enemy knowledge of friendly forces, while passive measures feed the enemy false or misleading information, fooling it into making faulty decisions.

8-72. Much of the counterreconnaissance effort is conducted by the cover group, which is likely built around reconnaissance or other light, mobile forces. These units conduct screening and covering actions throughout the frontal blocking zone. PLAA reconnaissance units are generally lighter and less powerful than their Western equivalents, and so they rely more heavily on deception and integrated firepower to achieve their objectives rather than direct action. Electromagnetic attack and defense must be carefully integrated into the counterreconnaissance mission, and information warfare may be used to deceive the enemy. Counterreconnaissance must be practiced throughout the depth of the battlefield, as the enemy has a variety of reconnaissance and intelligence capabilities capable of collecting at any point and in any set of conditions.

## BLOCKING ACTIONS

8-73. Blocking defensive actions are those actions that place a unit or group in a relatively static position, seeking to stop an enemy advance, attrit an enemy formation, or reduce enemy strength and cohesion through tenacious defending. Blocking actions are generally not decisive, but they are used to complement other actions on the battlefield by occupying enemy forces or slowing enemy advances.

8-74. Blocking actions make less use of depth than other defensive actions. Commanders executing a blocking action should seek to concentrate as much combat power as possible along the main line of defense, using terrain, obstacles, and entrenchments to enhance the strength of the position. Blocking actions are fundamentally less reliant on mobility and rapid reaction than other actions, and instead they rely heavily on terrain advantage, firepower, and brave defense to effectively resist the enemy. Commanders should generally consider blocking actions to be temporary in nature, as a static defense is always vulnerable to being overwhelmed, encircled, or isolated. Groups executing a blocking action should be prepared to retrograde if necessary, and commanders should make rear guards or reinforcements available if possible.

8-75. Blocking actions must make best use of firepower and obstacles to overcome the disadvantage of immobility. Artillery should be pre-registered and carefully integrated into the overall defensive plan. Firepower and obstacles should be employed cooperatively to create a protective boundary of firepower ambushes around vulnerable spots in the defensive line.

## REPOSITIONING ACTIONS

8-76. Repositioning actions are movements conducted to shift combat power during a defensive operation. These movements are intended to move combat power to blocking, counterattack, or fallback positions. As with blocking actions, commanders must use repositioning actions judiciously, as units on the move are more vulnerable to attack and firepower assault. Repositioning actions should be concealed if at all possible; movements at nighttime or during periods of bad weather are optimal. Repositioning actions should be performed rapidly with security provided to the main body, especially while it is in transit. Security must involve counterreconnaissance, counterfire, and screen or cover activity to counter direct action.

8-77. Repositioning actions are done for one of three reasons: to reposition a unit to a new, stronger, or more defensible blocking position; to attempt to shift the enemy's focus from one axis of defense to another; or to enable an adjacent unit to move, consolidate, or retreat. Repositioning actions are employed in careful coordination with blocking, screening, and covering actions to create an integrated, in-depth defensive effort. Commanders make use of repositioning actions to gracefully retrograde, ensuring that units on the move are well protected. Once a unit or group completes a repositioning action, it immediately converts to a strong defensive posture, enabling adjacent units or groups to conduct their own repositioning actions.

8-78. If the defense must retrograde into the deepest areas of the defensive zone, the defensive group must reposition its firepower and support units and its forward defensive groups. Commanders use the same principles when displacing rear area units as with forward units, while accounting for the former's lesser mobility and greater vulnerability. Air defense coverage becomes critical in rear areas, as rear area units on the move are highly vulnerable to air attack. Air defense coverage must be comprehensive and makes use of leapfrog or bounding overwatch techniques to maintain continuous coverage over moving units.

## BREAKOUT

8-79. Breakouts are conducted when a unit or group is encircled, flanked, or in danger of being encircled. The PLAA believes that most mechanized opponents will seek to encircle and isolate units or groups, using many of the same tactics and techniques the PLAA itself employs. This, in turn, has caused it to emphasize the importance of a unit being able to break out of an encircled or isolated situation while preserving manpower and avoiding destruction or surrender. Figure 8-16 on page 8-32 depicts a unit conducting a breakout action.

8-80. A breakout is as much a mental activity as it is a physical activity. Commanders must fight against the tendency of their troops to become passive, afraid, or despondent when facing long odds or a difficult situation. Encircled units will practically always be heavily outnumbered and outgunned. Through the use of skillful command, rapid maneuver, and careful deception, an encircled unit masses combat power at a weak spot in the enemy's position and, with the aid of other nearby units, groups, and systems, breaches the encirclement and moves quickly to a secure defensive position.

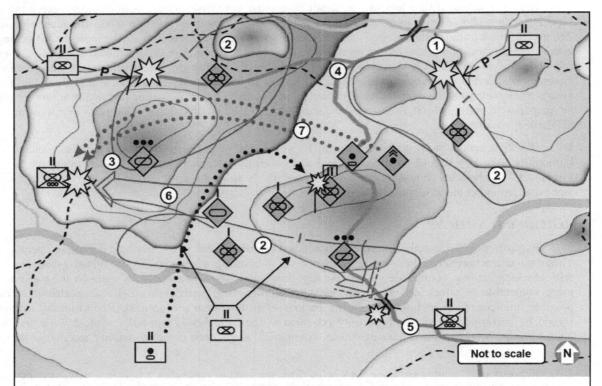

(1) A defensive group is isolated and surrounded by successful enemy penetrations. The group commander orders a breakout. (2) The defensive group establishes a strong defensive perimeter, using terrain and entrenchments to amplify combat power and enable units to endure enemy firepower and attacks. (3) Careful but rapid reconnaissance reveals that the enemy's weakest position is a light armored unit occupying a blocking position to the west. (4) The enemy likely anticipates a breakout to the northeast or the south, making use of main roads. Instead, the commander chooses a more challenging breakout route, hoping to surprise the enemy. (5) A reconnaissance platoon conducts a feint operation along an expected route in order to deceive the enemy. (6) An ad-hoc assault group comprised of the defensive group's reserves conducts a violent assault along the designated axis of advance. Though outnumbered, the group attempts to achieve localized superiority using deception and maneuver. (7) A firepower attack with the limited available assets supports the breakout assault. If the assault is successful, one company will conduct a rear-guard action while the rest of the defensive group conducts a withdrawal to a new defensive position.

| N | north | P | penetration |
|---|---|---|---|

**Figure 8-16. Breakout (example)**

8-81. Encircled units have likely endured heavy multi-domain firepower assaults and attacks through various flanks on the ground. As such, an encircled unit is likely significantly weakened and damaged, which presents a major challenge to the commander. The nature of the enemy's attacks will be unpredictable and constant, making extensive planning an impossibility. Preparations for a breakout are thus necessarily hasty, and commanders must assume a higher degree of risk when planning deployments and maneuvers.

8-82. The most critical element of a successful breakout is coordination with supporting groups and units—particularly firepower groups. This too presents a challenge, both to the encircled commander and to any supporting units. The enemy will certainly attempt to target and suppress any communications between an encircled unit and its potential support to ensure its isolation and preclude any attempts at relief. Commanders may employ creative or subtle methods for coordinating a breakout action. They may also encourage units to act independently. While this may decrease the concentration of combat power, it may enable a subordinate commander to effectively seize an opportunity and the initiative that otherwise passes the unit by.

8-83. The breakout action begins with rapid planning. The encircled commander should attempt to contact all subordinate groups and units to achieve some level of integration and coordination of purpose. Plans should be simple and easy to understand and transmit. Commanders at all levels must place a high premium on concealing their actual intentions, as they are likely under intense reconnaissance and surveillance. Deceiving the enemy about the time and place of the breakout effort is a key enabler.

8-84. The encircled commander's first objective is to build a highly defensible position. A strong defensive base helps to resist enemy penetrations and provides the anchor for an eventual breakout assault. Commanders take advantage of interior lines and strong terrain to create an in-depth, 360-degree defensive perimeter. Hasty fortifications and entrenchments should be constructed by all units and groups within the defensive area. The commander organizes the encircled units into two primary groups: a frontier group that forms the main defensive line, and a reserve or assault group that conducts the actual breakout attack.

8-85. Coordination with external support should be attempted and employed wherever possible. The coordination priority for an encircled unit is firepower. A firepower assault may disrupt, disable, or blunt the enemy's assaults and create useable breaches in the enemy's encirclement. Firepower coordination is very difficult for an encircled unit; however, and commanders should strive to keep firepower planning as simple as possible. Intricate firepower schemes are of little use; instead, commanders of assisting units should employ massed area fires in simple phased schemes, placing the onus on the encircled unit to make best use of the firepower assault. Adjacent assault, security, or reconnaissance units may also assist the encircled unit by conducting demonstrations, feints, or penetrations of the enemy's assault force. Coordination with these units is likely difficult or impossible, so much responsibility lies with the individual group commanders to exercise their discretion in assisting the encircled unit.

8-86. After encircled commanders establish a strong central defensive position, they gather combat power from the reserve group and establishes the axis of the breakout assault. Deception and concealment are paramount during this phase, as the enemy will attempt to discover the disposition of any breakout attempt and destroy it with firepower or ground assault. A breakout attack should take the form of multiple simultaneous penetrations, each targeting a weak spot along the enemy's assault force. These penetrations should be separate, but close enough to be mutually supporting. Commanders should strive to create a strong unity of purpose and strong morale in the assault group members. Though they are outnumbered, they can still achieve a victory and save their fellow soldiers if they succeed in breaking out of the encirclement. Once it is clear that one or more penetrations has succeeded, the commander should feed additional troops into the successful breach to secure it and enable the remainder of the encircled unit to withdraw.

8-87. As the remainder of the unit withdraws through the breach, the formerly encircled unit makes use of the reserve and assault groups, plus any outside assisting units, to secure the egress route. The enemy will react strongly to the breakout attempt, and it will attempt to counterattack. This counterattack must be anticipated and defeated with blocking actions and firepower. Movement out of the encirclement may be slow and arduous due to the condition of the encircled unit. Commanders must give these units as much time as possible to move to secure areas. The enemy will also enthusiastically pursue the formerly encircled unit with both air and ground forces. The higher echelon commander must establish a secure defensive position for the encircled unit to retrograde to, and commanders should launch spoiling attacks and employ air defense firepower to defeat the pursuit attack.

## WITHDRAWAL

8-88. A withdrawal is a carefully organized and planned retreat from a combat situation. There are two types of withdrawal: active and compelled. Active withdrawal is when a unit or group conducts a retrograde operation to take up a more defensible position, better defend an adjacent unit under assault, avoid enemy firepower, or lure the enemy into overextending itself. A compelled withdrawal is a retrograde operation undertaken in direct response to an enemy attack, when a unit is unable to continue resistance in its current position or it is unwise to do so.

8-89. Withdrawals are among the riskiest tactical actions. They are nearly always done while in direct contact with an enemy force, and they are often executed in close proximity to a superior enemy. Well-executed withdrawals integrate a wide array of combat power and a variety of methods to ensure the safety of the withdrawing unit and prevent the enemy from taking advantage of the potential disorganization and vulnerability in the defensive zone. Active withdrawals tend to be relatively well planned and carefully executed, while compelled withdrawals may require very fast planning and a great deal of flexibility in execution.

8-90. The prescribed order for withdrawal ensures that vulnerable friendly units are protected as the withdrawing unit breaks contact with the enemy. First, the cover and screen forces in the forward zones withdraw and move quickly to the established assembly point to the rear to establish a new security zone. Next, the unit's rear security and logistics elements move, passing lines with the cover and screen forces and establishing rear security and logistics support at the new assembly area. Next, the main body withdraws, along with the commander. This is considered the key point in a withdrawal operation, and it should be timed to take best advantage of weather, time of day, and other environmental factors that help conceal the main-body movement. This leaves only the rear-guard defensive element and the firepower element; these groups withdraw last. The firepower element moves quickly to prepared positions in the new defensive area. The rear-guard element conducts a stubborn retrograde action to delay and disrupt the enemy's pursuit attacks.

8-91. Once at the new assembly area, the defensive group rapidly establishes a new defensive area while simultaneously evacuating casualties, reorganizing subordinate groups, and establishing contact with any new adjacent units. Figure 8-17 depicts a unit conducting a withdrawal action.

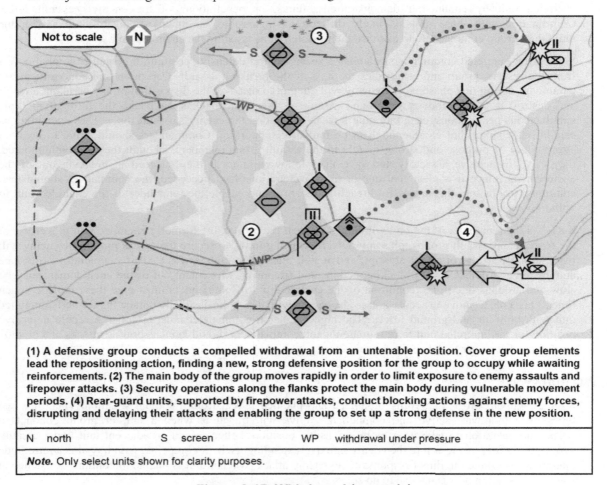

(1) A defensive group conducts a compelled withdrawal from an untenable position. Cover group elements lead the repositioning action, finding a new, strong defensive position for the group to occupy while awaiting reinforcements. (2) The main body of the group moves rapidly in order to limit exposure to enemy assaults and firepower attacks. (3) Security operations along the flanks protect the main body during vulnerable movement periods. (4) Rear-guard units, supported by firepower attacks, conduct blocking actions against enemy forces, disrupting and delaying their attacks and enabling the group to set up a strong defense in the new position.

| N | north | S | screen | WP | withdrawal under pressure |
|---|---|---|---|---|---|

*Note.* Only select units shown for clarity purposes.

**Figure 8-17. Withdrawal (example)**

## FIREPOWER DEFENSE

8-92. The PLAA considers the application of firepower to be the primary tool for defeating enemy ground assaults. Firepower defense is often the trigger for decisive counterattacks, and it has an important amplifying function for blocking and harassment missions. Firepower is employed throughout the depth of the battlefield and in all phases of the battle to disrupt the enemy, inflict casualties, defeat penetrations, and isolate enemy formations.

8-93. Firepower in the defense is provided by the firepower system, which is a mixture of mobile and static artillery, air defense systems, antitank and other direct fire systems, and attack aircraft. Firepower is organized both vertically and horizontally. Horizontal firepower seeks to dominate a specific axis—usually the main direction of defense—with a mixture of artillery, antitank, and mobile firepower. Vertical firepower seeks to control the sky above the battlefield with air defense firepower and make use of the air domain through fixed- and rotary-wing firepower systems. Firepower is concentrated against key points at key times, and it is carefully synchronized with maneuver to mass combat power. Firepower system effectiveness is enhanced when it is integrated with obstacles.

### Firepower Defense Purposes

8-94. The PLAA prescribes six primary reasons for employing firepower in the defense:
- Target enemy information networks and firepower systems.
- Strike assembling or unfolding troops.
- Resist enemy assaults.
- Defeat airborne, infiltration, and flanking maneuvers.
- Defeat penetrations.
- Target the enemy's depth.

These reasons are described in paragraphs 8-95 through 8-100.

#### Target Enemy Information Networks and Firepower Systems

8-95. A mixture of artillery, attack aviation, and electronic warfare targets the enemy's communications networks and artillery systems. This action seeks to either destroy or disrupt these key enemy systems, undermining the enemy's overall offensive action and enabling the decisive counterattack.

#### Strike Assembling or Deploying Troops

8-96. Enemy troops are at their most vulnerable when deploying and massing for an attack. (The preferred PLAA term is "unfolding" to mean deploying, but this publication uses deploying to reduce confusion.) Commanders should seek to target enemy forces during this phase of their offensive action by judiciously employing blocking forces and carefully timing firepower assaults. A mixture of aerial firepower and rocket and tube artillery is best suited for this firepower mission.

#### Resist Enemy Assaults

8-97. Directly supporting friendly units that are resisting enemy assaults throughout the frontier defense zone is the backbone of the firepower defense mission. These attacks seek to destroy or disrupt enemy formations before and during contact and provide agile, adaptive support to defensive groups as the battle evolves. This firepower is provided mainly by massed rocket and tube artillery.

#### Defeat Airborne, Infiltration, and Flanking Maneuvers

8-98. If mobile reserve units are unavailable, firepower may be used to disrupt or blunt enemy infiltrations and flanking maneuvers. These enemy units are often exposed, unarmored, or otherwise vulnerable, and targeting them with massed firepower may defeat them without the need for ground intervention.

### Defeat Penetrations

8-99. As the enemy's assaults begin to penetrate through the frontier defense zone, enemy units will encounter continued resistance and counterattacks from the depth defense zone. Firepower defense is employed throughout the depth defense zone to support counterattacks and cause casualties in enemy assault groups. Much of this firepower is provided by antitank and mobile groups.

### Target the Enemy's Depth

8-100. As the defensive action proceeds, the enemy's rear area is an important target. Enemy reinforcements, resupply efforts, fallback positions, command and communication nodes, and security forces are all commonly located in the enemy's depth area. A mixture of long-range rocket artillery and fixed-wing aircraft provides most of the firepower for this mission.

## Firepower Defense Capabilities

8-101. The bulk of the firepower system is comprised of four capabilities: artillery, antitank, aerial, and air defense. These capabilities are described in paragraphs 8-102 through 8-110.

### Artillery Firepower

8-102. Artillery firepower consists of indirect firepower systems that deliver either precision or massed fire throughout the depth of the battlefield. This includes rocket artillery, static and mobile tube artillery, and mounted and dismounted mortars. Artillery systems are the backbone of the firepower system. The PLAA envisions a high density of artillery systems to be a decisive capability in most defensive actions. Artillery systems perform every mission inherent in a firepower defense mission, including ground targeting, counterreconnaissance, antitank, information attack, counterfire, and air defense via targeting enemy aircraft support systems and aircraft on the ground.

8-103. If employed intelligently, the wide variety of artillery system capabilities creates a combined-arms effect, amplifying the effectiveness of each system while offsetting its vulnerabilities or limitations. Long-range rocket and static tube systems target enemy firepower systems, protecting other friendly firepower systems from enemy counterfire. They also target the enemy throughout its depth, disrupting operations and retarding penetrations. Mobile artillery systems respond rapidly to enemy breakthroughs, defending the heavy, less-mobile systems from enemy ground assault. Mobile mortars are used to flexibly enhance localized firepower in direct support of ground units, enabling heavier and more destructive systems to target the enemy's vulnerable rear areas. Artillery firepower thus creates interlocking rings of firepower systems that are weighted toward the main direction of defense.

### Antitank Firepower

8-104. The deployment of antitank firepower is a critical component of any defensive action against an opponent equipped with armored or mechanized forces. The PLAA employs an extensive array of antitank weapons, including assault guns, antitank guided missiles, short-range antitank weapons, and specialized artillery munitions. The PLAA employs a comparatively low concentration of antitank guided missiles, relying more heavily on massed short-range antitank weapons than its Western counterparts.

8-105. The employment of antitank firepower is one of the few ways in which the PLAA seeks to confront an enemy armored force. Commanders are encouraged to mass as much antitank firepower as possible along the expected avenues of approach for enemy armored forces and integrate the antitank groups carefully with obstacles and artillery. Concentrated antitank firepower seeks to destroy as many enemy armored vehicles as possible, with the understanding that the antitank group may suffer heavy casualties in the process. However, without this concentration of firepower and its associated risks, the enemy's armored thrust would likely succeed. Commanders must be aware of the short range of most PLAA antitank weapons, and they should seek to engage enemy armor at the closest range possible to maximize their effects.

8-106. Antitank teams must make best use of terrain, entrenchments, and concealment to survive the enemy's armored thrust. Though they are attacking the enemy head on, antitank teams should seek to attack individual vehicles or small units from the flanks or rear to strike at the weakest portion of the enemy's armor. Antitank teams must be protected against the enemy's inevitable counterstrike, which will employ a mixture of infantry and artillery. Careful concealment, fortifications, and security support offset the enemy's attempts to suppress the antitank group. Tactics such as ambushes and feints can multiply the effectiveness of the antitank group.

### Aerial Firepower

8-107. Aerial firepower is the combined capabilities of fixed-wing attack aircraft and attack helicopters. These two capabilities are somewhat different but largely complementary. Fixed-wing aircraft can deliver huge destructive power, but they are typically not well-coordinated with the ground element in contact, and they are often somewhat limited in their precision and accuracy. Thus, fixed-wing attacks are best employed targeting enemy rear areas: halting reinforcements, destroying supplies, and disrupting command and communication. Conversely, attack helicopter assaults are carefully coordinated with ground forces, and they are precise in their application of firepower. They are, however, far more vulnerable to ground and air defense firepower. Helicopters are best employed as a highly mobile reserve—targeting enemy flanking actions, encirclements, and infiltrations, as these are least likely to have substantial air defense capabilities or defensive counterair support. Commanders may also choose to employ attack helicopters to blunt the enemy's decisive attack. While this approach adds to the value of the decisive counterattack, it exposes the helicopters to more-concentrated air defense firepower.

### Air Defense Firepower

8-108. The PLAA is keenly aware of the massive destructive potential present in enemy air attack, and it places heavy emphasis on defending ground units from enemy aircraft. Air defense firepower is deployed throughout the defensive area using a tiered and layered approach, and it is enhanced by more-powerful, longer-range systems that are deployed across the theater. The PLAA employs a wide variety of air defense systems. Man-portable air defense systems (MANPADS) are widely fielded across formations and groups of all types. Self-propelled antiaircraft guns (SPAAGs) are widely employed by mechanized and motorized units, and short- and medium-range missile systems are present at the CA-BDE level. These systems are employed using a combined arms approach, wherein the capabilities of one system offset the vulnerabilities of another.

8-109. The PLAA divides target types by altitude. Ultralow-altitude air defense is provided by a combination of small arms, crew-served weapons, and SPAAGs. Low and medium air defense is provided by a mixture of SPAAGs, MANPADS, and missiles. High-altitude air defense is provided by theater assets, such as long-range missiles and fighter aircraft. Tactical echelons are most concerned with ultralow- and low-altitude threats, such as unmanned aircraft (UA) and helicopters, and seek to defeat these systems through a mixture of concentrated firepower that defends key assets and air defense ambushes that strike at enemy aircraft as they move along air routes. Air defense systems are oriented by section, while the coverage of the defensive zone is ideally all-directional.

8-110. Commanders develop the air defense firepower plan by designating assets to be defended, then building an air defense group. Commanders must strike the right balance between point defense of targets and active, offensive-minded air defense activities, such as ambushes. The PLAA seeks to field such a heavy weight of air defense firepower that the enemy is deterred from even attempting to operate aircraft in the defensive zone. If this deterrence fails, the air defense group typically has highly permissive rules of engagement-which may preclude friendly aircraft from entering the defensive zone. This risk profile is acceptable, as the PLAA believes the enemy to be far more reliant on air power than the PLAA is. The PLAA would prefer to have no aircraft from either side be present over the defensive zone than have aircraft from both sides.

## DEPTH DEFENSIVE ACTIONS

8-111. Depth defensive actions are those that typically take place in the depth defense zone. These actions are typically calculated to be decisive, as this is where the enemy's assault force is often the most vulnerable and friendly defensive measures are strongest. Depth defensive actions seek to definitively turn back the enemy assault or—ideally—annihilate the enemy's assault force. These actions are built around the depth defensive group, which typically contains a high proportion of firepower, mobile, or armored forces able to move quickly and deliver decisive attacks to the flanks or front of an enemy assault force. This effort is aided by comprehensive security throughout the depth defense zone, spoiling attacks, and blocking actions.

## Depth Security

8-112. Security throughout the depth defense zone may be provided by a number of different units or groups that either deliberately or dynamically conduct security operations. The cover group, having withdrawn from the frontal blocking zone, may be called upon to screen or cover the depth group. Additional security can be provided by the reserve group, scouting groups, or elements of the frontier defense group. It is critical that the commander make best use of available forces to maximize security. This will often involve reorganizing or reconstituting disparate groups into useful security forces in an ad hoc manner. Deliberate security measures may also be employed in the depth defense zone, but planning must account for the unpredictability of the enemy's axis of advance into the zone. If the enemy approaches from an unanticipated direction, security deployments may be useless unless they can be rapidly adapted.

8-113. Security forces may also take on the mission of shaping the enemy's actions and disposition in anticipation of the decisive counterattack. Shaping and deception operations attempt to orient the enemy's attention away from the axis of the main assault while fixing, deceiving, or blocking enemy forces that might react to the counterattack. Effects of demonstrations and feints are amplified by information warfare operations.

8-114. Security operations in the depth defense zone are broadly similar to those in other areas. Commanders of security groups, however, should be encouraged to defend more stubbornly and attack more aggressively than commanders of other groups, as the enemy should be prevented from entering the rear defense zone if at all possible. This may involve security groups being ordered to perform impromptu or hasty blocking actions, making best use of terrain, firepower, and hasty fortifications to prevent enemy penetration into rear areas.

## Spoiling Attack

8-115. Spoiling attacks are small-scale offensive actions that target an enemy formation's cohesion and morale. In the depth defense zone, spoiling attacks are primarily intended to defeat, blunt, or disrupt enemy maneuvers, such as flanking movements and penetrations, that may affect or threaten the primary counterattack effort. Spoiling attacks may be conducted by any type of group, but commanders should seek to avoid siphoning off power from the depth group for nondecisive actions, if at all possible. Spoiling attacks are discussed in more detail in paragraph 8-29.

## Counterattack

8-116. A counterattack is typically the decisive action within the depth defense zone. Ideally, it consists of an aggressive, concentrated assault by the depth defense group against a vulnerable portion of the enemy's assault force. The counterattack should be heavily supported by firepower, information warfare, and adjacent or supporting units. The counterattack seeks to break the enemy's cohesion and will, ending its thrust or penetration, and forcing it to choose between retreat and annihilation. Counterattacks are discussed in more detail in paragraphs 8-37 through 8-42.

## REAR DEFENSE

8-117. The rear defense zone is the deepest and most vulnerable section of the defensive battlefield. Commanders and defensive groups should strive to prevent significant enemy penetrations into the rear defense zone. Decisive actions should be planned for more-forward areas. Rear defense operations seek to blunt, delay, or defeat enemy penetrations, protect critical assets in the rear area, and—if necessary— enable the retreat or retrograde of vulnerable forces in the rear defense zone. Operations in the rear defense zone are undertaken by the reserve group, by rear security units—possibly including militia or paramilitary security units—and by any other groups that have moved into the rear area due to the maneuvers of the battle.

8-118. The rear defense zone typically houses important and potentially vulnerable systems and units, such as artillery groups, air defense groups, supply and logistics groups, command groups, and aviation groups. These groups are necessarily more vulnerable to direct attack. Protecting them or enabling their withdrawal is a critical mission in the rear defense zone.

8-119. Rear defense also includes the mission and associated tasks of defeating smaller enemy penetrations and infiltrations. The enemy seeks to establish a combat presence throughout the depth of the battlefield, and it will use aerial insertions and ground infiltrations to move small, mobile units into the rear area. These enemy units are charged with conducting covert direct operations, such as sabotage and raids, along with reconnaissance and surveillance activities. Rear defense units must be capable of detecting and neutralizing these enemy forces, either through direct action or deception. Commanders must assess the threat level in rear areas and ensure that the rear defense or reserve groups have the necessary combat power available to counter these threats. Figure 8-18 depicts defense operations in the rear defense zone.

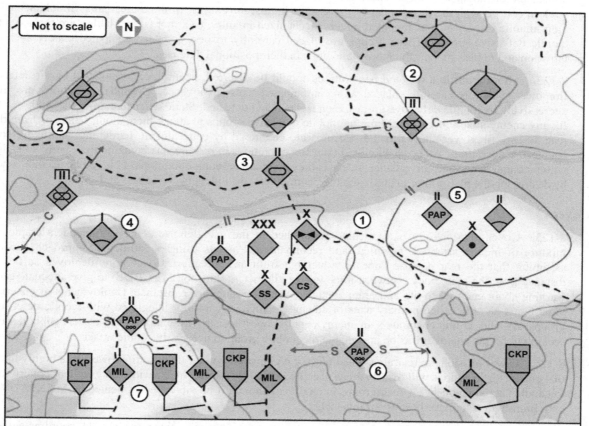

(1) A rear-area security zone is established around a group army headquarters element, combat aviation brigade headquarters, combat support and service support brigades, an HQ-9 SAM battalion, and an artillery brigade. (2) The cover group deploys reconnaissance units and two defensive groups to the frontal blocking zone, creating the main line of defense against enemy penetrations. (3) A heavy CA-BN serves as the depth group and combat reserve. (4) Medium-range HQ-16 SAM batteries are deployed through the frontal blocking zone, integrated digitally with the HQ-9 battalion. (5) Dismounted PAP battalions serve as security groups for the high-value targets. (6) Mounted PAP battalions conduct screen missions across southern avenues of approach, deterring infiltrations, raids, or surveillance by enemy units. (7) China militia units establish checkpoints along supply routes, conducting traffic and local route security.

| | | | | | | |
|---|---|---|---|---|---|---|
| C | cover | MIL | militia | S | screen |
| CA-BN | combined arms battalion | N | north | SAM | surface-to-air missile |
| CKP | checkpoint | PAP | People's Armed Police | SS | service support |
| CS | combat support | | | | |

**Figure 8-18. Rear area defense (example)**

## Rear Security

8-120.  Security in rear areas is a critical enabler of the overall defense. Numerous key groups and capabilities reside in the rear defense zone, and the enemy will seek to target these high-value assets with a mixture of infiltration, penetration—particularly by aerial forces—firepower, and electromagnetic attack. Security forces in rear areas consist of a mixture of static groups that resolutely defend the most valuable assets and mobile forces that react rapidly and aggressively to any intrusions.

8-121.  Commanders determine the responsibilities of the rear security groups according to the priorities of the battle and available forces. First, the commander must establish a clear priority of rear area assets. Those assets most critical to the overall prosecution of the defensive battle are placed highest. Then, a rear defense zone plan is developed which assigns available troops to defend or secure assets based on priority. Static or less-mobile groups conduct positional defensive actions. This cohort consists of militia, armed police, and dismounted soldiers. These groups carefully entrench around their specified defended asset and work in concert with the commander of the asset to harden the entirety of the critical area. Simultaneously, more-mobile forces, such as mounted police, mechanized infantry, or armor (if available), are formed into mobile defensive groups. These groups are placed carefully to best counter anticipated enemy axes of intrusion and to best support the more-static units in their positional defenses.

8-122.  Airborne infiltrations and penetrations should be deterred, neutralized, or attritted by air defense firepower which is carefully placed along anticipated axes of advance. Ground infiltrations are met aggressively by mobile defensive groups, then isolated and defeated. Static defensive groups should seek to fix any enemy attack forces, enabling the mobile defensive groups to attack them along flanks or from the rear. Enemy firepower attacks in the rear area are countered by concealment, hardening, and aggressive counterfire operations. A key enabler of counterfire operations in rear areas is breaking the enemy's linkage with reconnaissance and surveillance assets as they conduct targeting operations. This can be accomplished by destroying the platform or neutralizing the communications link between sensor and shooter.

## Cover

8-123.  Cover missions are of particular importance in the rear defense zone. Commanders employ a mixture of mobile armored, mounted, and reconnaissance assets using a 360-degree approach, always assuming that the rear defense zone is vulnerable from any direction. Cover units deploy a sufficient distance from the main body (between three to five kilometers [km] for a CA-BDE) to give adequate early warning about enemy activity and grant adequate space for stubborn resistance. Deployments of covering units focus on anticipated primary axes of advance. Units are carefully integrated with mobile defensive groups. As a cover unit encounters an enemy penetration, it resists while falling back to defensible terrain, seeking to fix the enemy intrusion and enable the mobile group to decisively counterattack.

8-124.  Cover missions in rear areas must place a high priority on air defense firepower. It is highly likely that enemy air intrusions will overfly areas occupied by cover forces; this provides an excellent opportunity to conduct aerial ambushes. Cover groups in rear areas should, if possible, be augmented by MANPADS or anti-aircraft guns to make best use of this tactic.

8-125.  If units are adjacent to one another, commanders between the units should coordinate their respective rear area cover missions. It may be unnecessary to deploy a cover force facing a friendly unit. Commanders must be aware of what is to the left and right of their rear defense zone and weigh their deployments accordingly. Commanders must also deploy cover forces along any seams or borders between units that may be exploited by enemy infiltration or penetration.

## Sabotage and Countermobility

8-126.  As rear defense groups may be both outmanned and outgunned in the event of a major enemy penetration, rear defense zone operations must make best use of both countermobility and sabotage missions. Countermobility efforts focus on building integrated obstacle systems that slow the enemy, force it to spread combat power or change axis of advance, or canalize it into ambush or counterattack zones. Rear defense groups typically have more time and assets to build obstacle systems, and they should rely heavily upon them to multiply available combat power. Particular attention should be paid to slowing the enemy as it enters engagement areas for firepower systems—both direct and indirect—and manipulating enemy movement to make it more susceptible to spoiling attacks or counterattacks.

8-127.    Sabotage is the deliberate, covert targeting and destruction of enemy assets to negatively affect the enemy's operations. While sabotage has its place everywhere on the battlefield, its usefulness is most pronounced in the rear defense zone. Militia, SOF, or paramilitary units are highly effective sabotage groups, as they can blend into the surrounding environment and subtly sabotage critical assets without provoking a strong enemy response. Sabotage efforts work in concert with countermobility efforts to slow the enemy advance through the rear defense zone. These efforts include, but are not limited to destroying bridges, roads, tunnels, or other mobility corridors; ambushing enemy supply elements as they move to resupply the enemy attack group; or conducting small ambushes of isolated enemy units. Sabotage teams should be kept small to minimize their signature, and sabotage efforts should be conducted in a decentralized manner.

This page intentionally left blank.

# Chapter 9

# Antiterrorism and Stability Actions

Antiterrorism and stability actions are conducted by a wide variety of different organizations within the Chinese government, ranging from both the regular and special operations (SOF) forces of the People's Liberation Army (PLA), to the People's Armed Police (PAP), to national and local law enforcement. These actions include domestic law enforcement, stability operations, maritime security, and international stability operations. The PLA broadly categorizes antiterrorism actions as part of maintaining both domestic and international stability, while PAP stability actions are largely domestic in nature and focus on internal threats and dissent. This chapter discusses national-level antiterrorism and stability operations, focusing predominantly on the mission types undertaken by People's Liberation Army Army (PLAA) and PAP units under the control and direction of the national government.

## OVERVIEW OF PLAA ANTITERRORISM OPERATIONS

9-1. Terrorism in the modern sense is a relatively new phenomenon in China, and the Chinese government is still working to establish a clear understanding of what constitutes both domestic and international acts of terror. The PLA defines terrorism as acts of violence that disrupt national unity and societal stability by creating casualties or damaging property. This definition is very broad, far more sweeping and ambiguous—possibly deliberately so—than most Western definitions. This helps to explain why the PLA views antiterrorist actions as supporting national stability. Instead of only acts of violence perpetrated to achieve a political or social objective, the Chinese definition also includes acts that threaten peace and stability within the country. This, in turn, gives Chinese authorities a very broad set of circumstances in which antiterrorism laws and operations can be applied. The growing Chinese interest in antiterrorism operations coincides with rising religious and sectarian tensions in specific regions of China and the rise of global terrorism through the latter half of the 20th century.

## THE INFORMATIONIZED BATTLEFIELD AND ANTITERRORISM OPERATIONS

9-2. The PLA recognizes several characteristics that shape antiterrorism activities: urgency, complexity, significant effects, joint nature, and asymmetry. As with all other military operations, China is acutely aware of the changes the informationized battlefield has brought to its antiterrorism mission. Unlike offensive and defensive actions, however, many of the effects the informationized battlefield has on antiterrorism operations are related to media, public opinion, public policy, and public perception.

### URGENCY

9-3. Antiterrorism missions are urgent in nature. Generally speaking, all of the initiative during a terrorist attack or campaign rests with the terrorist organization. Proactive antiterrorism efforts are possible, but they are limited in scope and effectiveness by the difficulties associated with identifying and neutralizing potential terrorists before they strike. This makes most antiterrorism activities reactionary and highly time sensitive. Constant vigilance is an important consideration for any unit or organization charged with conducting an antiterrorism mission, and rapid response to a terrorist event is one of the most important elements in preventing casualties, maintaining stability, and suppressing further terrorist acts.

## COMPLEXITY

9-4.    Antiterrorism missions are complex. Terrorist activities are often sophisticated and complex, making use of media, religious or political factionalism, civilians, and international actors to manipulate the situation to the terrorists' advantage. By framing an activity as defending religious freedom or democracy, terrorists can gain sympathy from other parts of the world. By coercing local civilians, they can conceal their activities and disrupt efforts to find and suppress them. This requires antiterrorism forces to be equally sophisticated and highly aware of both the effects their actions have on the global political dialogue and how they can push back against attempts to manipulate local civilian populations by building trust and maintaining a strong visible presence.

## SIGNIFICANT EFFECTS

9-5.    The effects of terrorism can be significant. One of the primary features of terrorism is that the psychological and political effects of terrorist acts can often far exceed the actual physical effects. Terrorism is designed to amplify the effects of violence by disrupting lives, bringing anxiety to a society, and pressuring those conducting antiterrorism efforts into overreacting to real or perceived threats. These effects may be realized even when a terrorist operation fails. The simple threat may be enough to achieve the desired effect. Antiterrorism efforts must be valued according to their ability to maintain the long-term peace, prosperity, and stability of a society, not just in terms of their ability to directly root out and defeat terrorist elements.

## JOINT NATURE

9-6.    Antiterrorism operations must be joint in nature. Antiterrorism missions require cooperation between different elements, perhaps more than for any other security activity. The complexity of terrorist activities, along with the fact that they often cross international borders, causes a high premium to be placed on open and effective cooperation. In this context, joint refers to three different forms of cooperation: between the military, police, and civilian antiterrorism units domestically; between the political and military parts of the government; and between different international actors.

## ASYMMETRY

9-7.    Terrorist actions are nearly always asymmetric. Terrorists seldom enjoy any significant parity in equipment or training to that of their opponents, and they seek to offset these disadvantages by attacking vulnerable targets from unforeseen directions. Most antiterrorist organizations are also asymmetric in nature. While they may be formally trained and equipped, most are drawn from police or militia organizations, and so they do not habitually employ aggressive military tactics. Operational methods used by both sides are described as irregular. Terrorists employ ambushes, raids, and sabotage, while antiterrorists employ their own ambushes, along with careful counterattacks, search and annihilate, and infiltration methods.

# PRINCIPLES OF ANTITERRORISM OPERATIONS

9-8.    China recognizes the significant challenges that antiterrorism operations face and their significant differences from more-traditional military operations. As a result, the principles that the PLA and the rest of the Chinese Security Apparatus apply to antiterrorism operations are unique and much more subtle than those that apply to more-symmetric kinds of actions.

## ACTIVE PREVENTION

9-9.    One of the most insidious elements of terrorism is that terrorist actions can still achieve much or all of their desired effects even when the mission itself fails. Terrorist acts can also target a virtually infinite number of different elements of a society, from the military, to education, to infrastructure, to government, to simple random targets. The only sure way to eliminate the terrorist threat is to suppress it when it is still in the embryonic stage. This requires antiterrorism forces to take constant, targeted preventative measures in order to identify and suppress terrorism before it reaches a more formal stage. Authorities must also take measures to reduce the intensity of terrorist actions by hardening vulnerable targets, keeping the local population informed, and ensuring that emergency response capabilities are robust and well maintained.

### RAPID REACTION

9-10. In spite of the best efforts to actively prevent them, terrorism events typically occur suddenly; often with little or no widespread warning. This requires the forces and organizations that respond to them to plan carefully for many different contingencies and react with great rapidity when a terrorist attack does occur. These organizations should be localized and know the population, terrain, and other characteristics of their assigned area. They must war-game many different possible courses of action to save time when actual events occur. When an event does occur, the planning is rapidly revised to meet the needs of the real-world situation. The priorities in these situations are (1) to secure the immediate area, (2) to reduce or contain civilian casualties, and (3) to rapidly counterattack and attempt to kill or capture the terrorist actors.

### JOINT ACTIONS

9-11. The complex legal framework of antiterrorism activities requires a unique joint approach to antiterrorism operations. PLAA, PAP, China Militia, national police, and local police must all work jointly under a unified command structure in order to achieve the best results. Because terrorism activities can take on an incredibly wide variety of different forms, structures, and capabilities, antiterrorism forces must be able to adopt any potential posture, from passive and cooperative to high intensity. This approach requires a true joint command that has the authority to unify and coordinate actions across numerous different organizations. This principle is inherently political, and deciding who actually commands an antiterrorism group is likely to be a decision made at higher levels of government. Leaders of subordinate organizations must put aside their own agendas and those of their commands in order to achieve harmony and unity of purpose within a joint group.

### LEGALITY

9-12. One of the most important elements of a terrorist action is the undermining of the rule of law and the customs and observances of a people. Terrorists deliberately attack these important institutions to magnify the effects of their violent actions by eliciting reactions of anger and fear from their victims. Antiterrorist groups must take great care to act in a legal manner, observing all relevant laws and regulations to make clear to the population that they have the moral high ground. Similarly, antiterrorism groups must carefully observe all important local customs and culture to gain the trust of the people. Acting in a legal and moral manner may put antiterrorism groups at a disadvantage versus wily and subversive opponents, but commanders must stick fast to this principle, even if it means incurring greater casualties or reducing the overall effectiveness of the antiterrorism operation. While it may seem prudent in the short term to shelve laws and decency in order to defeat the terrorist threat, behaving in a violent, illegal, or insensitive way toward the local population helps to breed future terrorists and may cost commanders the trust of the people they are trying to protect.

## ANTITERRORISM MISSIONS AND TACTICS

9-13. Unlike most other military operations, antiterrorism operations are prescribed to be tightly controlled and commanded by a central authority. This ensures that the antiterrorism operation is tightly integrated and that subordinate groups conduct their own operations in accordance with the principles of antiterrorism operations. Freedom of action by lower echelon groups is viewed as less critical, as the scope of antiterrorism operations is typically much smaller than major military operations, and the consequences for a subordinate's mistake are often more significant.

9-14. There are five primary antiterrorism missions; each has numerous subtypes and permutations based on the operational environment and the nature of the threat. Many of the same terms, techniques, and tactics carry over from full-scale military operations.

## RIOT CONTROL

9-15. Riot control is seen as one of the most important elements in maintaining domestic stability. Rioting is viewed as a direct threat to Chinese national security, and control efforts are centered on ensuring the safety of both people and property. The methods used, however, are aggressive—employing many of the same tactics and techniques used during high-intensity military operations. Riot-control operations consist of four phases: control and blocking; divide and encircle; pursue and destroy; and eliminate enemy remnants and suppress follow-on rioting. Figures 9-1 through 9-3 on pages 9-4 through 9-6 depict a riot control operation.

> *Note.* PLA doctrine does not make much differentiation between demonstrations and large-scale destructive riots. The methods prescribed are essentially high-intensity military tactics. It may be inferred that riot control in this context is referring to a large-scale insurrection or highly destructive riot. It should also be noted that all rioters are referred to as terrorists, independent of actual affiliation with a terrorist group or its motivations.

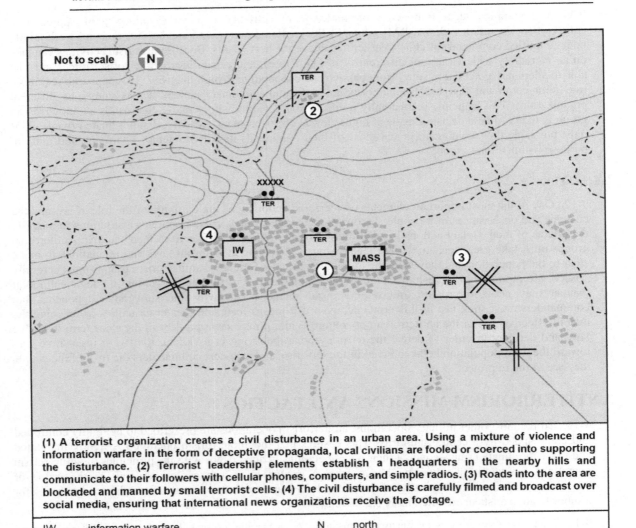

(1) A terrorist organization creates a civil disturbance in an urban area. Using a mixture of violence and information warfare in the form of deceptive propaganda, local civilians are fooled or coerced into supporting the disturbance. (2) Terrorist leadership elements establish a headquarters in the nearby hills and communicate to their followers with cellular phones, computers, and simple radios. (3) Roads into the area are blockaded and manned by small terrorist cells. (4) The civil disturbance is carefully filmed and broadcast over social media, ensuring that international news organizations receive the footage.

| | | | |
|---|---|---|---|
| IW | information warfare | N | north |
| MASS | mass demonstration | TER | terrorist |

**Figure 9-1. Riot control (example; part 1 of 3)**

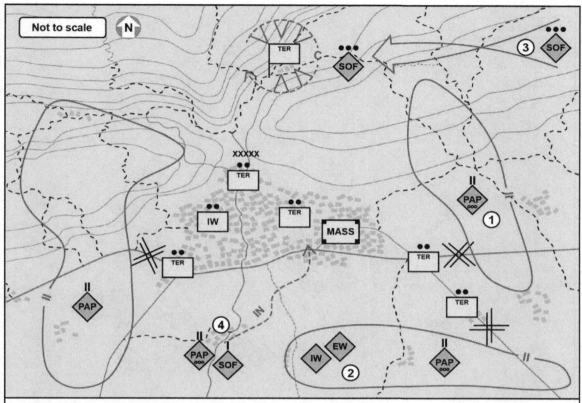

A PAP mobile detachment deploys to control the riot. (1) Three PAP battalions deploy to blocking positions around the area, cordoning it off and ensuring no travel in or out. (2) An EW group deploys to the area, suppressing all over-the-air communications, including cellular phones and radios—isolating distant terrorist cells from one another. An IW group begins targeting terrorist networks and computers and ensures that the area is cut off from internet access. (3) A SOF platoon, receiving a mix of electronic intelligence information and satellite imagery, identifies the location of the terrorist leadership cell. It conducts an air insertion and establish a cordon around the cell, isolating it physically from the rest of the terrorist cells. (4) A PAP battalion and SOF company comprise the assault group; they begin their intensive reconnaissance of the main riot area.

| | | | | | |
|---|---|---|---|---|---|
| C | cordon | IW | information warfare | PAP | People's Armed Police |
| EW | electronic warfare | MASS | mass demonstration | SOF | special operations forces |
| IN | infiltration | N | north | TER | terrorist |

**Figure 9-2. Riot control (example; part 2 of 3)**

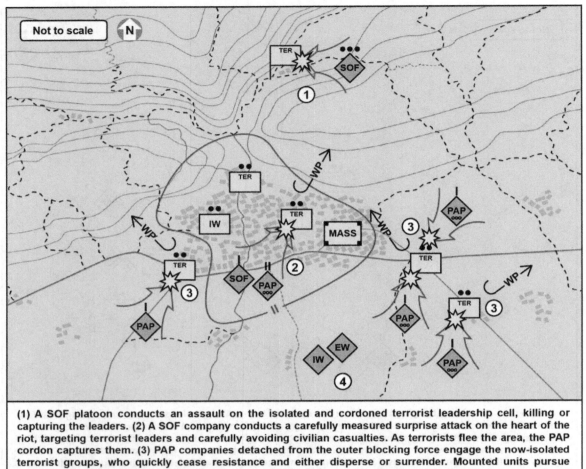

(1) A SOF platoon conducts an assault on the isolated and cordoned terrorist leadership cell, killing or capturing the leaders. (2) A SOF company conducts a carefully measured surprise attack on the heart of the riot, targeting terrorist leaders and carefully avoiding civilian casualties. As terrorists flee the area, the PAP cordon captures them. (3) PAP companies detached from the outer blocking force engage the now-isolated terrorist groups, who quickly cease resistance and either disperse or surrender. Mounted units pursue terrorists as they attempt to escape the area. (4) The EW and IW groups continue to suppress terrorist communications, while ensuring that the operation is recorded and rebroadcast as an antiterrorist propaganda tool, dissuading future riots.

| EW | electronic warfare | N | north | TER | terrorist |
|----|-------------------|---|-------|-----|-----------|
| IW | information warfare | PAP | People's Armed Police | WP | withdrawal under pressure |
| MASS | mass demonstration | SOF | special operations forces | | |

*Note.* Only select units shown for clarity purposes.

**Figure 9-3. Riot control (example; part 3 of 3)**

## Control and Blocking

9-16. The control and blocking phase always initiates a riot-control action. This phase consists of blocking groups who focus on denying possible rioter escape or reinforcement routes deploying around the riot area. Blocking operations should include both military and police forces, ideally working in concert with one another to properly leverage the capabilities of each. Blocking should also include the denial of communications and the suppression of ports, subterranean facilities, and other possible ways of escape or reinforcement. Blocking is continuous throughout the operation, but blocking forces should maintain a flexible approach to their deployment, ready to reinforce key areas if and when they come under direct assault. During this phase, comprehensive reconnaissance is conducted throughout the riot area to ascertain the positions of the leaders, key rioter defensive positions, and any other important pieces of intelligence to inform the upcoming decisive phases.

### Divide and Encircle

9-17. The divide and encircle phase is the decisive phase of the riot-control operation. It employs a mixture of surprise attacks and storming attacks to overwhelm the rioters, with a focus on dividing the various groups, isolating them from their leadership, and then annihilating them. Some groups are highly vulnerable to isolation and will cease resistance once cut off from their leaders, while others will fight determinedly—even when completely cut off and surrounded. Antiterrorist forces must be aware of the tendencies of their opponents as they face destruction. Some riot participants will surrender willingly and without further resistance, while others will respond with suicidal violence, even when facing certain defeat. Antiterrorist force leaders must employ force measures appropriate to their opponent. Less-fanatical adversaries may be willing to peacefully surrender if offered the opportunity, while suicidal, fanatical individuals pose a lethal threat until they are captured or killed.

### Pursue and Destroy

9-18. The next phase of the riot-control operation is the pursuit and destruction of any remaining remnants. It is critical that as many terrorists as possible be captured or killed, as any who escape will remain a threat and can easily consolidate and rally new recruits in a short period of time. As most riots occur in urban environments, pursuit will likely be on city roads and through populated areas. Pursuing forces integrate both land and air assets to keep careful track of the positions of those retreating, canalizing them into less densely populated areas and blocking their escape to safe havens. Once the groups are pushed to a favorable position, they are blocked from all sides and attacked until captured or destroyed. Antiterrorist groups may also employ techniques such as ambushes to surprise and defeat retreating rioters.

### Eliminate Enemy Remnants and Suppress Follow-on Rioting

9-19. Once most rioting forces are captured or killed, the operation transitions from pursuit to search. Antiterrorist groups conduct comprehensive searches of the combat area in order to find any remaining isolated remnants. This search can be either immediately after the successful destruction of the main group or some time afterward, giving remaining enemy forces time to relax and believe themselves safe from reprisal. Searches must be careful, leaving no blind spots or areas unchecked, but the antiterrorist group also must not disrupt the lives of peaceful citizens in the search area.

## ACTIVE ANTITERRORISM ATTACK

9-20. Active antiterrorist attacks are those attacks that proactively target terrorist bases of operations and leaders. They may be independent missions, or they may be conducted concurrently with or in support of other antiterrorist activities. The general approach to these attacks is much the same as for a riot-control or ground offensive operation. Decisive attacks penetrate the enemy defenses and isolate enemy units, allowing their destruction in detail. There are, however, two important activities that must be emphasized during a decisive pre-emptive attack. First, reconnaissance and search must be comprehensive and meticulous, ensuring that the entire battlefield is carefully assessed and all terrorist targets identified. This may be a lengthy process and commanders must be patient, not launching their attack until it is complete. Second, antiterrorist groups must emphasize the blocking actions or cordon that surrounds the terrorist target. Terrorists likely have preplanned escape routes in place. They will rapidly flee the area if threatened, so a substantial cordon force may be required in order to fully contain the area. A careful search should reveal all major possible egress routes.

## HOSTAGE RESCUE

9-21. Hostage rescue missions integrate multiple different competencies in such a way that decisive application of both physical and psychological force frees the hostages and eliminates the terrorist group. Hostage rescue missions are arguably the highest profile of all antiterrorism mission types. They often are carefully followed by the world at large, and they have outsized significance and consequences for both security forces and society at large.

9-22. A hostage rescue mission employs five different groups: a blocking group, an assault group, a reconnaissance group, a psychological warfare group, and a rescue group. These groups work in concert with one another to break terrorist resistance while protecting the hostages. The rescue mission commences after hostages are taken and antiterrorist forces are organized and deployed. The first phase of the operation is the unfold phase, where the blocking unit deploys around the periphery of the combat area and secures land, sea, and air routes from either ingress or egress. The other groups spend this period organizing themselves and developing the plan of action.

9-23. Psychological warfare operations commence in the next phase of the mission. This naturally centers on the psychological warfare group, although a variety of other organizations may have roles to play. The psychological warfare phase typically consists of negotiating with the terrorist group. Negotiations are never used to concede to the political demands of the terrorists, but rather to begin the process of isolating them, dividing their forces, and disintegrating their strength. Negotiations should be deliberately difficult, causing psychological stress to the terrorists, but they should not be so aggressive as to cause a threat to the hostages. They may also serve to distract terrorists from ongoing assault preparations, enabling the assault group and reconnaissance group to conduct their missions without detection. If it is possible to resolve the situation through negotiation without making political concessions to the terrorists, this should always be the first choice.

9-24. The final phase of the hostage rescue mission is the raid or attack. This consists of the reconnaissance and assault groups carefully assessing the situation, developing an assault plan, and then bravely and aggressively executing the plan. Reconnaissance efforts should ascertain the exact position of the terrorists and hostages, identify all of the potential ambush points or strongpoints in and around the objective, and provide the assault team with a comprehensive plan of attack against the heart of the terrorist group. The psychological warfare effort should have caused distress and division among the terrorists, possibly compromising their defenses and isolating the different terrorist groups. Careful concealment during the approach phase, coupled with surprise and precision in the attack, are the best ways to ensure that the terrorists are killed and the hostages unhurt. Nonlethal munitions—such as smoke, flash, and flare rounds— may be used to temporarily disorient the terrorists without harming hostages. Sniper teams should also be employed, both as reconnaissance platforms and as overwatch. Multiple sniper teams are deployed in a variety of different positions around the combat area, creating a variety of threats to the terrorists. Snipers may also be employed as ambush teams to engage terrorists attempting to escape the area.

## BORDER CONTROL AND SECURITY

9-25. China places a high priority on border security in general, and in particular where border regions are unstable or dangerous. Border security falls to a number of different organizations, from local and national police active units of the PLAA. The territory designated for enhanced security measures is called a border control area.

9-26. Borders cross three domains: land, sea, and air. In order to effectively control the border, the border-control force must control traffic through all three domains. Land borders are controlled by a mixture of firepower and physical occupation. Forces align along widely used corridors and critical areas, and they make use of mobile forces to react to unexpected intrusions through less-patrolled areas. Firepower may be used to augment these forces or to close areas of the border that are not physically manned. Border-control areas do not necessarily need to be placed directly on the border. If terrain or other conditions dictate, they may extend some distance into Chinese territory.

9-27. Border control areas must be carefully cordoned through the use of blocking forces and checkpoints. Blocking forces employ defense-in-depth techniques to ensure that the area is secure from both ingress and egress, and checkpoints control the flow of traffic in and out of the area. As border-control efforts become harsher, the negative effects on the civilian population and economy increase. Local officials must carefully balance the need for security with the negative effects on the lawful population and cross-border trade. Ports are another area of emphasis: while they are not likely to be targeted by offensive action, they provide ingress routes for undesirable forces and cargo. In a high-security border zone, inspection efforts through ports must be significantly enhanced.

9-28. Air and sea border control typically involve the use of exclusion zones to deny the use of airspace or waterways to unwanted traffic. A no-fly zone or restricted airspace denies opponents the use of airspace over the border-control zone, while a naval exclusion zone denies use of waterways. These measures must be widely publicized to ensure that domestic and international aircraft or ships do not inadvertently violate them, and sufficient air and naval and coast guard forces must be present to ensure that the measures can actually be enforced.

# THE INFORMATIONIZED BATTLEFIELD AND STABILITY AND SECURITY ACTIONS

9-29. Security and stability missions are among the highest priorities for the Chinese Security Apparatus. In past generations, much of the domestic security mission fell to the PLA. Today, most of the burden is shouldered by the PAP, along with a mixture of national and domestic law enforcement agencies. In addition to peacetime stability and security missions, the PAP has a significant role in supporting wartime missions. The PAP should not be underestimated as a military and paramilitary force; it is comparable to a regional power's national army in size and combat power. The PLA recognizes several characteristics of stability and security actions: political nature, hybrid warfare, threat strength, and international peacekeeping. These characteristics are discussed in paragraphs 9-32 through 9-35.

9-30. The PAP command structure and control mechanisms are still somewhat opaque to outsiders. It appears that PAP units are organized into brigade-sized formations called mobile detachments that are subsequently assigned to a specific region. In wartime, group armies may take over operational control of PAP units to support group army operations, primarily in rear area and security roles.

9-31. PAP equipment is discussed in the following appendixes:

- Appendix A, maneuver capabilities.
- Appendix B, fire support capabilities.
- Appendix C, air defense capabilities.
- Appendix D, aviation capabilities.
- Appendix E, engineer and chemical capabilities.
- Appendix F, network and communications capabilities.
- Appendix G, special operations forces capabilities.

Figure 9-4 depicts a genericized PAP mobile detachment as it might be organized for wartime support operations.

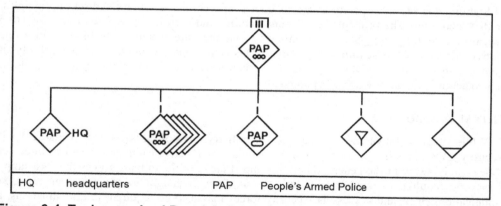

**Figure 9-4. Task-organized People's Armed Police mobile detachment (example)**

## POLITICAL NATURE

9-32. While all military activities are political to a certain extent, security and stability operations are more heavily influenced by politics than are force-against-force operations. Security and stability forces must coexist with local populations, and they may be required to protect them from attack or to suppress civil unrest. Social media is more readily available in security zones removed from the front line. Security-group activities are often high-profile and likely to be seen and publicized. As a result, security and stability forces are given more leeway and more responsibility for implementing domestic policy than other forces as they conduct traditional law enforcement alongside their military duties.

### HYBRID WARFARE

9-33. Hybrid warfare seeks to significantly influence behind-the-lines operations, where security and stability forces typically operate. Activities of a hybrid opponent may include the use of either civilians or soldiers in civilian attire to conduct raids, ambushes, or other direct actions in rear areas; the use of firepower or electromagnetic attack to target key assets in rear areas; the use of information attack to influence civilian populations throughout the security zone; or the use of special operations or irregular forces to conduct deep direct action throughout the security zone. The clash between security and stability forces and the enemy's offensive hybrid warfare elements will likely be a key factor on the informationized battlefield.

### THREAT STRENGTH

9-34. The ability to threaten rear areas with a variety of different attack methods has expanded as firepower and irregular threats have proliferated and grown in capability. Security areas now face many of the same levels of threats as forward defensive areas, but security and stability forces do not enjoy the same combat power as most of their military counterparts. Security forces must ensure they can effectively neutralize the anticipated threats throughout their assigned security area, and they must carefully allocate combat power to ensure the right balance of forces in accordance with their higher echelon's priorities.

### INTERNATIONAL PEACEKEEPING

9-35. The multiple disciplines required of stability and security forces place them in an ideal position to function as international peacekeepers. As China seeks to expand its international role through participation in peacekeeping missions, it is likely that the government will lean heavily on its PAP units to function in an expeditionary capacity. As with any other stability-type mission, peacekeeping missions are often complex and ambiguous, requiring mentally agile leaders and versatile units.

## PRINCIPLES OF STABILITY AND SECURITY OPERATIONS

9-36. Stability and security operations are conducted in accordance with a fixed set of principles, as are all other PLA activities. The principles governing stability and security operations, however, are referred to specifically as guiding thoughts, likely inferring that they are meant to be broader in scope and less restrictive than other Chinese military principles. This is in keeping with the view that stability and security missions are more complex and ambiguous than other operations, and thus they require greater latitude for leaders to act in accordance with their judgment.

### SITUATIONAL AWARENESS

9-37. While situational awareness is required for military units of all types, it is of particular importance for security units operating in concert with PLA units on an active campaign. Security force commanders must be acutely aware of the general objectives and strategy of the military forces they are supporting, and they may be required to operate with little or no direct oversight in securing rear areas, supply and communications lines, and key assets. Security forces may also be required, in dire situations, to face off against much more powerful opponents in order to delay the enemy or buy time for the supported force to reposition or retreat. Security forces must be prepared to make this sacrifice either when called upon or when they recognize the situation requires it.

### KEY-POINT CONCENTRATION

9-38. Security forces operate with far less density than do regular army units. Contiguous deployments throughout the security zone are likely impossible, considering available forces and the number of different assets that must be defended. Commanders must carefully prioritize what they want to defend and from what type of threat. This enables the security force to properly allocate its subordinate forces to concentrate greater security on key points throughout the security zone. At the same time, security forces must be able to continue operations even when enemy forces are moving through the security zone, provided that they are moving through unimportant or indefensible areas.

### DEFENSE FOCUS

9-39. Security forces will most likely be outnumbered when facing conventional opponents, and they may be outnumbered even by irregular opponents. This makes offensive actions difficult, as security forces will seldom be able to concentrate sufficient combat power to attack and destroy an opponent. Instead, they must focus on the defense, building their operations around steadfast protection of key points, enabled by extensive entrenchments and a defense-in-depth approach. The doctrine of active defense still applies; however, when a security unit can achieve local superiority, it must attack and spoil the enemy's preparations. Security units contribute significantly to the victory of the larger military force by delaying enemy forces, attritting enemy formations, and disaggregating enemy capabilities in the security zone, setting the enemy up for decisive defeat by a counterattack.

## STABILITY AND SECURITY MISSIONS IN SUPPORT OF JOINT MILITARY OPERATIONS

9-40. As with other stability and security missions, those performed in support of joint military operations are primarily defensive in nature. They can range across many combinations of domains.

### DEFENSIVE ACTIONS

9-41. The PLA recognizes that assets throughout rear areas, such as air and seaports, bridges, railheads, command and communication nodes, and missile operation sites will be high-value targets for the enemy. Much of the responsibility for defending these assets from land, sea, and air attack falls to security forces.

### GROUND DEFENSE

9-42. Ground defense actions by security forces employ a defense-in-depth approach to make it difficult or impossible for the enemy to successfully attack key assets. Enemy strike methods against rear areas include direct attack by ground forces, secret attacks by special or irregular forces, and firepower attacks. Security forces employ a mixture of blocking actions along key avenues of approach, strong defenses around key assets, and the use of ambushes and raids against enemy intrusions.

9-43. Blocking forces employ either static defense methods or checkpoints. Static defense—allowing no one to pass through the defended area—is employed when the chances of enemy action are high, and the movement of friendly forces or civilians through the area is not critical to wider operations. Checkpoints—allowing carefully screened personnel to pass through—are employed in more-secure areas or where friendly movement is required. Static defense techniques are more secure than checkpoints, but they create more hardship for the local population and friendly forces. Commanders must weigh the risk of threat to an asset with the needs of friendly forces and the local population when choosing which method to employ.

9-44. Security units should make extensive use of entrenchments, hardening, and fortifications to enhance the defensive effort. This is particularly true for static assets: if an asset cannot be moved, it is far easier for the enemy to target and destroy it. Hardening and fortifying these assets forces the enemy to commit more resources to their destruction while protecting security forces from enemy attack and multiplying the effects of their defensive efforts.

9-45. As enemy forces move into the security zone, security forces engage the intruding forces through a mixture of stubborn defenses, counterattacks, raids, and ambushes. If the enemy force is relatively small and weak, a decisive counterattack or ambush may seek to annihilate it or force its withdrawal. If the enemy force is powerful, security forces conduct stubborn retrograde operations, delaying the enemy's advance, inflicting casualties, and disrupting enemy cohesion in anticipation of a powerful counterattack from a main-body defensive group. Retrograde operations cease once the security group reaches a key asset. At this time, the security unit must bravely defend without further retreat, using all of its remaining combat power to resist the enemy's assault.

## COASTAL DEFENSE

9-46. Security-force responsibilities revolve primarily around neutralizing shore raids and harassments that target wide areas of shoreline. Most of these duties fall to the Chinese Coast Guard, a subsidiary of the PAP. Due to the vast distances of vulnerable coast in China, this necessarily requires security units to be widely dispersed, limiting their combat power. Security forces are not intended to resist major amphibious operations; this responsibility falls to the PLAA. Instead, they are to remain vigilant against small-scale enemy intrusions over the shore, mainly reconnaissance raids, sabotage raids, and insertion or extraction of enemy agents. They also take on numerous civil missions, including coastal patrol and policing, regulation and protection of trade, and interdiction of smuggling.

## AIR DEFENSE

9-47. Security-force responsibility for air defense is similar to coastal defense. Security forces have little to no active air defense capability, but they are instead responsible for responding to aerial insertions of ground forces into the security zone. China fears the use of helicopters to conduct air insertions of troops into rear areas, and it relies heavily on security forces to counter this threat. Security forces will be widely dispersed and cannot resist a major air assault operation, but they are expected to detect and neutralize smaller aerial insertions of commandos, raiders, or enemy agents. The best method is to ambush helicopters as they descend in altitude and consolidate. An alternative method is to target enemy forces while they are vulnerable immediately after landing. This requires security forces to detect and track enemy air elements as they move into the security zone, then rapidly mass combat power in and around the landing zone. Security units should make extensive use of observation posts integrated with radar and the wider air defense network.

## DEFENSE OF SOCIAL STABILITY

9-48. China places a high priority on security and stability at all times; this emphasis also extends to combat theaters. Security forces play an important role in maintaining social stability in rear areas. The policy focus is almost entirely on social stability in Chinese domestic territory, but this emphasis likely extends to any foreign territory that has been occupied, conquered, or annexed during the course of a conflict. There are three primary components to the defense of social stability mission: guard socially vital assets, maintain social order, and address the effects of enemy attacks.

### Guard Socially Vital Assets

9-49. A subcomponent of guarding key points throughout the security area, security forces have responsibility for guarding those assets that are vital to social stability. In a wartime theater, these assets include distribution stations for food, water, and fuel; civilian transportation nodes; civilian government facilities; and public utilities. Unlike more-aggressive security missions that may face enemy regular forces, security groups charged with guarding these assets should focus on reducing their visibility and either concealing the target or deceiving the enemy as their primary courses of action. The most likely enemy actions against targets of this type are irregular forces and recruited criminal elements operating in a secret or concealed fashion. As such, measures to guard these assets are as much police work as military security work, and integration between civilian police forces and security forces is vital.

### Maintain Social Order

9-50. The component of maintaining social order reflects the ongoing Chinese fear of counterrevolutionary activities that may leverage propaganda and politics to undermine the Chinese government and military, seeding discord through rear areas. This mission type necessarily involves interaction between security forces and the population at large as the former tries to locate and neutralize enemy or other elements attempting to sow discord, cause riots, harass military units, or sabotage infrastructure. Commanders must work together with civilian law enforcement units when conducting this mission, and they should carefully weigh the risk to friendly operations against the risk of alienating either friendly or occupied peaceful civilian populations through the use of excessive force.

### Address the Effects of Enemy Attacks

9-51. Enemy efforts in rear areas will sometimes succeed, whether they are from firepower attacks, ground attacks, or attacks by irregular forces. Security forces have an important role to play in dealing with the effects of these attacks. If the local population does not see national leaders reacting strongly and quickly to minimize the effects of a successful enemy attack, its morale will suffer. In particular, casualties suffered in rear areas must be quickly and efficiently cared for, either treated on site or evacuated to proper medical facilities. Destroyed buildings, roads, bridges, or other infrastructure must be first secured, then rehabilitated as quickly as possible not only for the strategic benefit, but also to show the local population the strength of the government. The primary role of security forces in this mission is care of casualties, followed by the securing of destroyed or damaged areas.

### Defense of Supply Routes and Traffic

9-52. The most important mission for security forces in the security area of an active combat theater is the management of military traffic and the defense of transportation and supply lines. The PLA rightly fears rear area attacks on its transportation infrastructure as a means to cutting off its powerful frontline army units. In areas not near the front line, the bulk of this mission falls to security units. There are three primary components to defending supply routes and traffic: route security, traffic management, and emergency repairs. Route security requires constant, active vigilance against a variety of enemy attack methods that might seek to destroy friendly convoys or bridges, tunnels, and other key transportation assets. Active defense is conducted along main routes to discover enemy activity and neutralize enemy efforts.

9-53. Traffic management involves the careful routing of military traffic to ensure that the most valuable convoys use the most secure routes, and that cargo is effectively prioritized in accordance with the needs of the theater. Civilian traffic must be carefully controlled to ensure that priority military traffic has the right of way. This is a specialized skill set, and security units must be prepared to conduct this mission at any time and place.

9-54. Emergency repairs are those construction activities that take place when a route has been successfully attacked. Security units must secure the immediate area, drive off any lingering enemy forces, and then protect the construction engineering teams as they quickly repair the damage to re-enable military traffic flow.

### SECURITY FOR THE MAIN FORCE

9-55. The PLA envisions its security forces in an expanded role, providing direct security for the PLAA. This likely involves defense of the rear and flanks of a main body, along with defending critical assets and routes of march within the main body's combat area. It also includes management of the military region, including the wider geographic area occupied by the main body. Security forces integrate with the reconnaissance and security elements of the main body when conducting this mission, with armed police supporting military operations by securing rear areas.

9-56. This mission requires the security force to carefully integrate with the main body, keeping close track of its activities. If the main body advances, the security force provides rear area security and route security. If the main body retreats, the security force may be required to provide flank security or, in dire situations, conduct rear-guard actions against enemy forces. Main-force security operations also contain a counterreconnaissance element. Security forces must ensure that enemy scouts cannot infiltrate rear areas and conduct reconnaissance or targeting of the main body or key assets in the security area. This may involve direct action against enemy scouting elements, some of which may be powerful armored units. Armed police units have only minimal antitank capabilities; these must be enhanced if the security unit anticipates contact with enemy heavy forces.

### DECEPTION ACTIONS

9-57. The PLA also envisions its security units contributing to ground actions through deception, primarily through the use of feint actions. Security forces—particularly armed police—are well suited to missions of this type due to their high density of light, mobile armored vehicles and the ability to move quickly through a variety of different terrain types. Five different forms of feint are envisioned for security forces:

- Feint movement: a security force moves in such a way that it deceives the enemy into thinking a larger army movement is occurring.
- Feint assembly: a security force simulates a more-powerful ground force assembling on the ground for an offensive action.
- Feint attack: a security force conducts an actual limited attack on an enemy position.
- Feint defense: a security force conducts a limited or simulated defense against an attacking enemy.
- Feint withdrawal: a security force pretends to be a larger force withdrawing from the battlefield.

9-58. All of these measures are supported with false intelligence, including fake radio traffic. Their general intent is to fool enemy reconnaissance and surveillance assets into thinking that the security unit is actually a mechanized or armored friendly unit and to manipulate them into acting a certain way due to this misbelief. This, in turn, puts actual main body forces on the offensive, knowing that the enemy has been fooled and is pursuing an action against a deception unit. Commanders must remember, however, that security forces are not equipped to operate in close quarters with enemy heavy forces, and they should strive to protect their assigned security forces during deception actions.

# Appendix A
# Maneuver Capabilities

## MANEUVER CAPABILITIES FUNCTIONAL OVERVIEW

A-1.  People's Liberation Army (PLA) maneuver forces include the People's Liberation Army Army's (PLAA's) combined arms brigades (CA-BDEs), special operations forces (SOF), army aviation brigades, and the handful of remaining infantry divisions, along with the airborne brigades of the People's Liberation Army Air Force (PLAAF) and the marine brigades of the people's Liberation Army Navy (PLAN). The paramilitary formations of the People's Liberation Army Reserve infantry units, People's Armed Police (PAP), and China Militia units may be used for rear area or flank security or economy of force and deception (feint) missions. While the entirety of the PLA is undergoing substantial reform and modernization, significant but decreasing numbers of older systems are expected to be fielded throughout frontline formations through 2035—the scheduled date for PLA equipment modernization to be complete. Unlike Western armies that tend to field very homogenous formations from an equipment perspective, the PLAA employs a very wide variety of vehicles with varying capabilities and readiness levels. CA-BDEs with more-critical missions or in more strategically important areas receive newer and better equipment, with older equipment gradually trickling down to less-important commands. The PAP employs a wide variety of equipment types, generally tending toward light, motorized forces along with inexpensive mortars, recoilless rifles, armored personnel carriers (APCs), and limited numbers of helicopters.

A-2.  The core of the PLAA's maneuver component consists of around 73 CA-BDEs of three different types: light (motorized), medium (mechanized), and heavy (armored), along with a handful of legacy division structures. In addition, there are approximately 18 divisions in the reserves and militia, though their equipment and readiness levels are much lower than those of the active-duty force. All or nearly all of the active component is either motorized or mechanized. CA-BDEs regularly conduct large-scale training maneuvers against a full-time opposing force brigade. The PLAA brigade structure is described in detail in chapter 2.

## MANEUVER CAPABILITIES EQUIPMENT OVERVIEW

A-3.  The PLAA employs a wide variety of tanks, infantry fighting vehicles (IFVs), and APCs, ranging from high-end modern systems to Cold-War relics. Historically, the PLAA has been reticent to throw anything away. As new equipment arrived, older equipment was handed down to lower-readiness units, then to reserves, then militia, and finally to massive storage facilities. The recent set of PLAA reforms, however, moves away from this model, seeking greater homogenization through a force equipped entirely with modern equipment. This necessarily reduces the overall size of the army, and it promises significant cost savings in the maintenance and operation of obsolete equipment. At the present time, however, the PLAA still employs a variety of armored vehicles, with most CA-BDEs still equipped with older or obsolescent vehicles. Due to the vast differences in capability between in-use systems, it is critical that PLAA units' equipment be carefully assessed on an individual basis, rather than generically.

---

*Note.* The equipment overview provided in this appendix is not intended to be exhaustive, but rather to introduce the most widely employed variants of major equipment types used by the PLA.

---

### TANKS

A-4.  The PLAA fields a massive number of tanks, most of which are main battle tanks (MBTs), with a handful of light and amphibious tanks. The MBT force currently has more than 6,500 vehicles—more than twice the number fielded by any other country. Many of these, however, are early Cold-War vintage and of dubious utility on the modern battlefield. Early Chinese tanks were either direct copies or heavily influenced by Soviet designs, but the more-modern tanks are entirely indigenous and show more design similarities with Western tanks than with Russian ones.

A-5.   Despite its age, the Type 59 tank is still fielded in limited numbers in active units, and large numbers are maintained in reserve or mothball status. The Type 59 was derived from the ubiquitous Soviet T-54/55 tank and shares most of its design features. Its armor is homogenous steel, and its early 100-mm gun is inadequate against modern MBTs. The PLAA has, however, periodically upgraded its Type 59 fleet, adding modern thermal sights, a larger gun, improved gun stabilizers, explosive-reactive armor (ERA), and improved crew survivability systems. Modernized Type 59 tanks are lightweight, reasonably mobile, possess relatively modern fire control systems, and are simple to operate and maintain. However, their protection—even when upgraded with ERA—is poor by modern standards. Though the tanks employ a copy of the once-ubiquitous L7 105-mm main gun that can fire both tungsten and uranium penetrators, main gun performance is poor compared to modern larger guns.

A-6.   The Type 80/Type 88 tank was the first Chinese indigenous tank design. The PLAA mixed Western and Soviet design philosophies in the Type 80, using a Soviet-style chassis and turret design with a Western gun and crew configuration. The result was an MBT that was competitive with the T-72, T-80, and early M1 Abrams designs. The first variants entered service in the early 1980s, and they have been continuously upgraded. Those currently in service are essentially comparable to the much-later Type 96. Upgrade versions feature advanced computerized fire control, a mixture of homogenous and applique composite armor, a 125 millimeter (mm) smoothbore gun capable of firing penetrators, high explosives (HE) and missiles, ERA, and a significantly upgraded engine. While an older design, upgraded Type 80/88 tanks are competitive with older Western MBTs, and they can threaten even the most modern designs.

A-7.   The Type 96 tank is the most widespread modern MBT in the PLAA inventory. A fully indigenous design derived from the Type 80, the Type 96 was the first Chinese tank to feature full composite armor and a Western-style arrow-shaped turret, in contrast to the bowl-shaped turrets of the Type 80/88. The Type 96 is a relatively light modern MBT with excellent mobility, leading the PLAA to deploy most of its Type 96s in hilly regions and other areas of difficult terrain. Composite ERA and laser jammers have been added to the original Type 96, improving its survivability against antitank guided missiles (ATGMs). The Type 96 was originally only an iterative upgrade, and it was not truly competitive with either its Western or Russian counterparts when it was first deployed. It was built in large numbers, however, and upgraded variants— including Types 96A1 and 96B—have improved lethality and protection.

A-8.   The Type 99 tank is China's newest MBT. It represents the PLAA's new commitment to modernization and investment in technology. Using the same design approach as the Type 96, the Type 99 employs a Russian-style chassis with a Western-style turret. It features a full suite of modern protective systems, including spaced composite armor enhanced with ERA; full chemical, biological, radiological, and nuclear protection; crew survivability upgrades; and an active protection system. Its 125-mm main gun can fire sabot, HE, and missiles, and its advanced suspension and powerful engine give it excellent cross-country mobility. An upgraded variant, the Type 99A, enhances the protection scheme with thicker armor and multi-layer ERA, a more powerful engine, and improved electronic warfare and network capabilities. Due to its high cost, it is not anticipated that the Type 99 will fully replace the Type 96, and the Type 99A will only see service in elite "digitized" units.

A-9.   The Type 62 is an evolution of a Cold-War tank that has been modified and modernized into a light tank platform. Developed specifically to operate in the challenging mountains and jungles of southern China, the Type 62 was essentially a scaled-down Type 59, armed with an 85-mm gun and a lightweight armor package. At around 30 metric tons, the Type 62 filled the gap between heavy MBTs and very light airborne and amphibious tanks. Modernized variants feature upgraded armor and targeting packages, but they are badly undergunned and have virtually no offensive capability against modern MBTs. These tanks are still in limited use in specific geographic areas, but they are being phased out in favor of the Type 15 light tank.

A-10. The Type 15 is a brand-new light tank developed to take over the role filled by the Type 62. It features the ubiquitous 105-mm rifled gun and composite armor, along with an advanced remote weapons station and autoloader. The Type 15 is likely not intended to compete directly with modern MBTs, but rather to enhance the firepower of mountain and jungle units. It is also likely tailored for the export market—numerous lower-tier militaries around the world are seeking to replace their Soviet-era equipment, and a modern tank in the 30-metric-ton class meets many of their requirements. The Type 15 was just recently accepted by the PLAA, and it is in only limited use by operational units.

A-11. The <u>Type 63</u> is a widely proliferated amphibious light tank. Early Type 63 models were built primarily to provide PLAA forces with an armored platform capable of crossing rivers and other small bodies of water. These vehicles are still in limited service in lowland regions. The upgraded Type 63A was built specifically for the People's Liberation Army Navy Marine Corps (PLANMC), and it is capable of crossing open-ocean areas. Both types are capable of firing while embarked, though the Type 63A has a greatly enhanced on-the-move capability. Both tanks have only light homogenous armor, and they are not intended to withstand attacks from modern ATGMs or enemy tanks. While still in use by both the PLAA and PLAN, Type 63s are being phased out in favor of more modern amphibious tank systems.

A-12. The <u>ZLT-05</u> amphibious tank is a hybrid IFV, tank, and assault-gun platform with impressive performance across even large bodies of water. A modern 105-mm gun employs sabot, HE, and ATGMs from land or water, using a modern fire control computer and extensive stabilization. While heavier than the Type 63, the ZLT-05 still features only light armor, and it is intended to protect only against fragmentation and light weapons. The ZLT-05 is intended to be the standard amphibious tank throughout the PLANMC and, as such, it is the backbone of the PLA's short-range amphibious capability.

## INFANTRY FIGHTING VEHICLES

A-13. Despite being one of the early adopters of IFVs, the PLAA employs a lower density of them than do other large militaries, particularly as compared to Russia. There are numerous varieties and configurations of Chinese IFVs, and categorizing them can be complex: weight, configuration, and firepower are not conclusive, as they are with the U.S. military. Instead, one must look at the organization of the unit in question: a mechanized (medium) CA-BDE employs its armored vehicles as both IFVs and APCs, while motorized (light) CA-BDEs employ their vehicles almost exclusively as APCs. Chinese IFV designs have generally mirrored Soviet and Russian designs going back to the BMP-1 of the 1960s, although PLAA preferences now seem to be trending toward favoring wheeled as opposed to tracked vehicles. The PLAA actively fields roughly 4,000 IFVs of all types, plus a large number of older vehicles in reserve status.

A-14. The <u>Type 86</u> IFV is a copy of the ubiquitous BMP-1, and it is still widely fielded by active PLAA formations. The original Type 86 featured a 100-mm gun, while the newer Type 86A was rebuilt with a 30-mm autocannon. Variations include troop transport (with an eight-soldier capacity), scouting, engineering, ATGM carrier, air defense, amphibious, and command vehicles. The Type 86 is tracked, with good cross- country mobility and considerable firepower for its weight. Its protection is poor, however, consisting only of thin homogenous steel and minimal survivability improvements.

A-15. The <u>ZBD-04</u> IFV is a heavy tracked vehicle that strongly resembles the Russian BMP-3. It features a powerful engine, improved protection, and impressive firepower, though it weighs nearly twice as much as the Types 86 and 86A that it was built to replace. The ZBD-04 is broadly similar to the M2 Bradley in configuration, size, and capability. It employs both a high-velocity 100-mm main gun and a 30-mm chain gun, coupled with ATGMs and machine guns—a remarkable amount of firepower for an IFV platform. It can carry seven soldiers, and it is fully amphibious, though it cannot cross ocean waters. Its armor is modular—ranging from the homogenous steel hull to applique composite and ERA add-ons—giving it frontal protection from cannon fire in addition to all-around machine gun and fragmentation protection. Other variants of the ZBD- 04 include tank destroyer (replaces passengers with additional missiles), heavy tank destroyer (replaces normal HJ-8 ATGMs with heavy HJ-10 ATGMs), command, and reconnaissance vehicles. The reconnaissance variant is the heaviest and most-powerful scouting platform in the PLAA. An improved variant, the ZBD-04A, adds additional armor protection and improved fire control and network capability, though at the cost of additional weight.

A-16. The <u>ZBD-03</u> IFV is the PLAAF's primary airborne armored platform. The ZBD-03 is very small compared to its contemporaries—it is only eight metric tons, as compared to 24-plus metric tons for the ZBD-04 and nearly 30 metric tons for the M2 Bradley. Despite its small size, it enjoys excellent firepower. A 30-mm autocannon and a lightweight ATGM launcher give the ZBD-03 punch similar to much-larger

IFVs. Its protection, however, is very light, and it is likely only fully resistant to small arms and fragmentation. The IFV's mobility is outstanding, thanks to its light weight, high power-to-weight ratio, and advanced suspension. The ZBD-03 is both air mobile and air droppable, and it forms the primary armored component of the PLAAF's airborne corps. It is envisioned as the main offensive capability of the airborne corps, and it is intended to take on enemy rear area forces, garrison forces, and other soft targets as part of airborne operations. This IFV was not designed to confront heavy mechanized forces. Variants include a command vehicle and a tank destroyer variant with a larger ATGM launcher instead of a gun.

A-17. The ZBD-05 IFV is the PLANMC's amphibious fighting vehicle. It shares a chassis and machinery with the ZLT-05 light amphibious tank, the primary difference being that the IFV carries eight passengers and employs a 30-mm autocannon, rather than the 105-mm main gun. The ZBD-05 can cross open ocean with relative speed and safety, and it can engage targets while embarked due to advanced stabilization of the gun and missile launchers. The ZBD-05's protection is similar to that of the ZLT-05. It is necessarily light and only proof against small arms and fragmentation. The ZBD-05 is in wide service with the PLAA's amphibious brigades and the PLANMC; proliferation will likely increase significantly as the PLANMC expansion continues.

A-18. The ZBL-08 is an IFV built on the relatively new Type 08 8 x 8 wheeled vehicle chassis. The Type 08 chassis is broadly similar to the U.S. Stryker or the Russian BTR, though it is an indigenous Chinese design, and it is somewhat heavier than either of its contemporaries. The ZBL-08 variant carries seven passengers, and it employs a 30-mm autocannon plus ATGMs. It has good cross-country mobility, and it is notably fast on improved roads, making it a first choice for urban combat situations. Its armor is likely proof against heavy machine-gun fire and fragmentation, though it may have limited protection against light cannon from some directions. The IFV variants are employed in much the same way as their tracked counterparts—as firepower support for dismounted infantry. A reconnaissance variant is equipped with a sophisticated sensor suite including radar, thermal sensors, electro-optical sensors, and a laser designator, in addition to the gun and missile package of the IFV.

## ARMORED PERSONNEL CARRIERS

A-19. As with IFVs, the PLAA employs a relatively low density of APCs as compared to other major militaries. APCs comprise the main mobility element of light CA-BDEs, providing light infantry with enhanced mobility and protection but contributing only limited offensive firepower. Most PLAA APCs are tracked, but a growing number are 8 x 8 wheeled vehicles. In addition to the PLAA's APC force, the PAP employs a significant number of APCs in support of its security, domestic stability, and antiterrorism roles.

A-20. The Type 63 APC is one of the first Chinese indigenous armored fighting vehicle (AFV) designs, and it is very similar to the U.S. M113 in size, capability, and role. Simple, lightweight, and amphibious, it has enjoyed a long and useful lifespan—just like the M113. It employs lightweight steel armor to defend against small arms and fragmentation, and it can carry up to 10 passengers. The Type 63's design has lent itself to a number of different variants, including command vehicle, ambulance, and mortar or artillery carrier. It has been periodically upgraded with more-powerful engines and better suspension, and it will continue in PLAA service for the foreseeable future.

A-21. The ZSD-89 APC was developed as an offshoot of an export product. The PLAA was impressed by its carrying capacity and mobility and ordered it to serve alongside the Type 63s already in service. The ZSD-89 is one of the world's largest APCs in terms of capacity, able to carry 13 passengers or a large number of ATGMs, artillery rounds, or mines in other configurations. It is amphibious and protected against fragmentation and heavy machine-gun rounds. Like the Type 63, the ZSD-89 has been developed into a large number of different variants, including mortar and artillery carriers, command vehicles, scouting vehicles, supply and recovery vehicles, tank destroyers, and IFVs.

A-22. The Type 08 is the APC variant of the ZBL-08 8 x 8 wheeled AFV. It can carry up to 10 passengers, and it enjoys a substantial advantage in both speed and range as compared to tracked vehicles. The Type 08 employs a number of advanced features, such as a remote weapons station and advanced suspension, and it is likely to be the PLAA's APC of choice as older vehicles are retired, particularly in the medium CA-BDE. The Type 08's chassis is large and well-suited to a variety of roles, including command, communication, artillery and mortar carrier, antiaircraft gun, reconnaissance, ambulance, electronic warfare, engineering support, and assault gun. The PAP also employs Type 08s in domestic security and antiterrorism roles. This is the largest and most-powerful vehicle employed by the PAP.

A-23. The ZSL-92 is a light APC that employs two different 6 x 6 wheeled chassis that are collectively referred to as the WZ-551. Despite a long and troubled development period, the ZSL-92 is widely proliferated in the PLAA, the PAP, and internationally. An IFV variant is equipped with a 25-mm autocannon, and it can carry nine passengers, while the APC variant employs a heavy machine gun, and it can carry up to 11 passengers. Due to its light armor, however, it is unlikely that the PLAA employs the ZSL-92 in an offensive role. The ZSL-92's light weight and road mobility make it a natural choice for the PAP, and large numbers are employed in engineering, crowd control, and security roles.

## TANK DESTROYERS, ASSAULT GUNS, AND ANTITANK GUIDED MISSILE VEHICLES

A-24. The PLAA places a very high priority on antitank firepower in tactical formations. As a result, it extensively modifies virtually all of its AFV platforms to enhance their antitank, antipersonnel, and anti-obstacle firepower. The two most common approaches to these modifications are either to replace a machine gun or cannon with a tank gun (usually in the 105-mm class), or to add additional ATGM launchers and replace the crew compartment with additional ATGM storage. These systems provide enhanced firepower to light and medium CA-BDEs, typically operating as one company within a battalion.

A-25. Tank destroyers and assault guns are generally similar in capability and appearance. Nearly all employ an APC or IFV chassis, with the crew compartment occupied by the machinery and ammunition to support a 105-mm main gun and turret. Nearly all of the PLAA's 105-mm guns are derived from the widely proliferated British L7 gun of the 1960s. These guns still pose a threat to even modern armor because of upgrades to their muzzle velocity and ammunition. In addition to their mobile antitank role, these guns also have useful antipersonnel and anti-obstacle capabilities, and they are employed by battalion or group commanders whenever enhanced firepower is required. Among the most common assault-gun types are the ZTL-11 (a modified Type 08 APC) and the PTL-02, based on the ZSL-92 APC.

A-26. ATGM vehicles are widespread in PLAA formations of all types. Virtually every battalion-level organization employs at least a platoon of ATGM vehicles. For lighter units, this represents the bulk of their mobile antitank firepower. Due to their thin armor, ATGM vehicles are best employed in ambushes or standoff attacks where they cannot be targeted by enemy tanks or IFVs. They are almost exclusively a defensive weapon, though they can employ their missiles to target bunkers or other fortifications if required. Virtually every PLAA APC model has been modified to be an ATGM vehicle.

## LIGHT ARMORED VEHICLES AND WHEELED VEHICLES

A-27. Most of the PLAA's light tactical vehicles are in the Mengshi family, a group of vehicles derived from the U.S. H1 Hummer, which is the civilian version of the U.S. M998 high mobility multipurpose wheeled vehicle. Like the M998, the Mengshi family of vehicles has been periodically upgraded, starting primarily as soft-sided utility vehicles and now featuring extensive armor, mine-resistant chassis, and remote weapons stations. These vehicles fill a wide variety of roles, from simple transport, to ambulance, to command, and to maintenance, fulfilling practically every role available for a light wheeled tactical truck. The PLAA, carefully following coalition activities in Iraq and Afghanistan, also developed the Mengshi into capable patrol vehicles, scouting platforms, and local fire support in the form of the CSK series of vehicles. The latest Mengshi vehicles are significantly larger than their Hummer predecessors, strongly resembling U.S. joint light tactical vehicles in both size and capability. These vehicles are ubiquitous throughout the PLAA and the PAP, providing much of the motorization support for ground forces along with a growing protection and firepower suite.

A-28. The PLA employs a variety of heavy wheeled vehicles, varying widely in size and payload. The MV3 is the generic medium truck, usually operated in a 6 x 6 configuration. It is largely comparable to the U.S. M939 or light medium tactical vehicle, though it is a newer design than either. Numerous different models of larger trucks provide progressively greater payload. The largest varieties are up to 10 x 10 in configuration and serve as transporter-erector-launchers for intercontinental ballistic missiles (ICBMs) and long-range surface-to-air missile (SAM) systems. The density of cargo trucks—particularly the medium-lift cargo trucks that are the backbone of tactical logistics—is relatively low in the PLA, and it affects China's ability to conduct expeditionary operations or operations a significant distance from its internal logistics structure.

## SMALL ARMS AND DIRECT FIRE WEAPONS

A-29. China's military independence from the Soviet Union can be tracked with the growth of its small arms industry. While the early PLA relied entirely upon Soviet and Soviet-derived weapons, years of investment built the Chinese small-arms industry into one of the world's largest and most sophisticated. This development was paralleled by the advancement of the PLAA's service rifle: over the last 50 years it transitioned from a simple copy of the Soviet SKS, to the development of variants based on the AK 47, AKM, and AK-74, to completely indigenous contemporary designs.

A-30. The QBZ-95 is the PLAA's primary service rifle today. A bullpup design, the QBZ-95 is optimized for ease of use and volume of fire. It is well regarded as a simple, sturdy, and comfortable weapon. Its 5.8-mm x 42-mm round offers more power and better ballistic performance than the standard 5.56-mm North Atlantic Treaty Organization (NATO) round, though at slightly increased weight and recoil. It is less accurate than most of its Western competitors, and it lacks the modular upgrade capabilities—such as rail systems— enjoyed by most modern assault rifles. The next generation of the QBZ-95 is thought to include a flat-top rail system instead of simple iron sights. The QBZ-95 has been modified into grenade launcher, carbine, and light support-weapon variants. Older service rifles—mostly AKM and AK-74 variants—are still in wide use across militia and reserve formations.

A-31. Like most Western Armies, PLAA formations employ three classes of machine gun: squad automatic weapons, primarily the QBB-95 light machine-gun variant of the QBZ-95; general purpose machine guns— a mixture of older Soviet-derived weapons and newer Western-style general-purpose machine guns; and heavy machine guns in the 12.7-mm caliber class, including a direct copy of the U.S. M2HB. In addition, 14.5-mm heavy machine guns are employed as antiaircraft weapons throughout tactical formations. General- purpose machine guns are operated by machine-gun teams and mounted on vehicles of all types; heavy machine guns are typically mounted on armored vehicles. Heavy antiaircraft machine guns are fielded in each combined arms battalion (CA-BN) heavy-weapon company. Generally speaking, PLAA machine guns can be considered comparable—and in many cases practically identical—to their Western equivalents.

## MAN-PORTABLE MISSILES AND ROCKETS

A-32. The PLAA employs a very high density of man-portable antitank weapons throughout its formations, echoing a general emphasis on antitank capability that is visible throughout the organization. The PLAA's historical approach to man-portable antitank warfare was simple mass, equipping formations of all types with huge numbers of unguided rockets—mostly based on the ubiquitous Soviet RPG-7—and simple ATGMs, predominantly the HJ-73, a copy of the Soviet AT-3 Sagger. These systems are still in use and still pose a threat to many armored vehicles due to periodic upgrades and the density of their deployment. The PLAA, however, is beginning to take a quality-over-quantity approach with regard to its ATGMs, seen most easily in the development and fielding of the HJ-12. Developed specifically to compete with the U.S. FGM-148 Javelin, the HJ-12 employs most of the latest ATGM developments, including a top-down attack profile, tandem warheads, soft launch, and a fire-and-forget targeting capability. These systems are expensive and complex, however, and their density is limited by their cost.

A-33. The PLAA still employs a high density of unguided man-portable rockets throughout its formations, particularly in light infantry units. These weapons employ simple HE, antitank, fuel and air explosive, incendiary, and demolition warheads, filling a variety of different roles and providing significant enhancement to the firepower of the infantry squad. The PF-89 is the most widely proliferated shoulder-fired antitank rocket system, and it is very similar to the NATO AT-4 in both size and capability. The FHJ-84 is a lightweight twin- rocket system that fires an incendiary warhead for antipersonnel or demolition use, and the PF-97 is a heavy fuel and air explosive rocket system. The PF-98 does not have a Western equivalent. It is a powerful 120-mm antitank and antipersonnel rocket system, usually fired from a bipod. This weapon gives the light infantry squad a lethal capability against a modern MBT, though its weight, size, and cost limit its proliferation.

# MANEUVER CAPABILITIES AND LIMITATIONS

A-34. In just over 20 years, the PLAA's maneuver capabilities evolved from a force mixture of light infantry conscripts and obsolete tanks to a nearly fully motorized and mechanized force that employs a variety of cutting-edge armored vehicles and advanced guided munitions. This evolution is incomplete, however. A significant number of CA-BDEs are still equipped with older systems, and they are likely to remain so for the foreseeable future. So too has the movement towards professionalization within the maneuver force produced mixed results. While the quality of both recruits and noncommissioned officers (NCOs) has increased significantly, the PLAA still struggles with decentralizing planning and operations at lower tactical echelons. The maneuver force has also struggled with the complexities of the informationized battlefield and modern mechanized forces. The technical demands of the informationized soldier are far greater than for conscripts of past generations, and the maintenance and logistics demands of mechanized forces are much greater than those of light infantry forces.

A-35. The most significant capability enhancement in the PLA is the development of a viable expeditionary capability. Previous PLA forces planned only to operate in Chinese territory, in neighboring territories, and in Chinese territorial waters. In order to establish itself as a great power, however, China believes that a powerful ground expeditionary capability is required. This is the primary driver behind the expansion of the PLANMC. It is likely the most significant effort in the PLA to meet the new expeditionary requirements. Despite these efforts, the PLA still cannot deploy or sustain a heavy mechanized force outside of shared land borders and its territorial waters, limiting the combat power of an expeditionary force.

A-36. PLAA leaders at tactical echelons are technically and tactically competent, with fairly regular training in both staff and command functions. The officer corps is professionalized and well educated, with many officers completing both advanced degrees and advanced military educations. Many PLAA officers are commissioned enlisted personnel. The PLAA relies very heavily on its officer corps for leadership functions. A professionalized NCO corps is a relatively new development. PLAA NCOs were historically just conscripts with a few years' experience, but starting in the late 1990s the PLAA began exploring the idea of career NCOs. This cohort is still underdeveloped compared to the U.S. Army, with minimal staff experience and no command experience, but the PLAA's belief in this new system is reflected in decreasing officer recruitment and the continued expansion of the NCO corps.

A-37. Despite these reforms and a greater emphasis on independence and initiative at lower tactical echelons, PLAA formations still struggle with independent action. Small unit leaders are afraid to make decisions due to a zero-tolerance culture, and officers remain hesitant to empower subordinates lest they underperform and thus reflect poorly on the officer. Except in elite units such as SOF or reconnaissance, the battalion is likely to be the lowest echelon capable of operating independently for any significant period of time, and even the battalion's independence is limited by a small staff and a low density of support assets. Significant resources are being devoted to mitigating these gaps, however, such as the establishment of training facilities that mimic U.S. Army Combat Training Centers and the development of a professionalized, dedicated opposing force.

A-38. The Chinese soldier is tough, physically fit, and deeply loyal to the country and the Communist Party of China (CPC). China's growing urbanization and improving social infrastructure are increasing the health, education, and general quality of its conscripts across the board. Training quality, however, varies wildly. Some units receive substantial training resources, such as maneuver exercises against a professional opposing force, while others may not even receive adequate ammunition for regular marksmanship practice. It is challenging for outsiders to effectively assess a unit's training readiness level due to the wide variations between brigades, even within the same group armies. Readiness levels vary even more among reserve and militia units. Some maintain regular training schedules and musters, while others exist practically in name only. Reservists are nearly all taken from former active-duty soldiers and, as such, enjoy a baseline level of training, but a reserve or militia unit, if activated, will take an extended period to reach a useful training readiness level. Conversely, the elite elements of the PLA—SOF, prestigious CA-BDEs, airborne, and marine forces, among others—enjoy substantial training resources and are likely comparable in competence to well-trained Western formations.

This page intentionally left blank.

**Appendix B**

# Fire Support Capabilities

## FIRE SUPPORT FUNCTIONAL OVERVIEW

B-1. No other capability area within the People's Liberation Army Army (PLAA) has received greater emphasis during the recent period of reform than fire support. In just over two decades, PLAA fire support evolved from a collection of aging Soviet-derived equipment to a largely indigenous, sophisticated, widely varied, and numerous collection of gun and rocket systems. The PLAA employs a mixture of mortars, howitzers, and rockets as the backbone of its fire support, including a notably high density of modernized rocket systems capable of employing precision and near-precision guided munitions.

B-2. The PLAA's emphasis on firepower manifests most clearly in its investment in modernizing its fire support capabilities. Experiences and observation led PLAA leaders to conclude that their maneuver units faced significant challenges when matching up against their approximate equivalents in the U.S. and Russian armies, so the decision was made to pursue enhanced firepower as a means of offsetting that capability gap. Joint fire support—from fixed-wing or rotary-wing aircraft, or from ships or submarines—is much less well developed. Close air support is virtually nonexistent as a capability, and the PLAA employs a very low density of attack helicopters, though this number is set to increase. Naval firepower is similarly largely absent from the People's Liberation Army's (PLA's) portfolio, though amphibious forces can provide much of their own firepower support through amphibious tanks, assault guns, and mortars.

B-3. The PLAA fields ground-based fire support in very high densities throughout its formations. Battalions of all types operate organic medium mortars, while combined arms brigades (CA-BDEs) employ heavy mortars, howitzers, and rocket artillery (multiple rocket launchers—MRLs) in their organic fire support battalions. The group army's artillery brigade also features a mixture of guns and rockets, adding in a heavy rocket battalion. These units may be retained by the group army to mass fires on a critical target, or they may be task-organized down to the CA-BDE in order to enhance the CA-BDE's firepower. The CA-BDE's organic firepower is roughly equivalent to that of a U.S. brigade combat team (BCT): 27 guns, along with a light rocket battery. Most of the CA-BDE's guns are 122 millimeter (mm) versus the predominantly 155-mm guns of the BCTs, but the BCT lacks an organic rocket capability. The CA-BDE possesses a higher density of guns and better mobility of firepower systems, while the BCT enjoys the greater range and destructive power of larger, higher-caliber gun systems. The density and destructive firepower of reinforcing fires is strongly in the PLAA's favor. The artillery brigade includes up to 72 guns, plus up to 40 MRLs of various caliber. These systems give the group army a significant advantage in both range and firepower over equivalent Western formations.

B-4. Fire direction, targeting, and forward observation have also been major areas of investment for the PLAA. China is the world leader in unmanned aircraft system production and development, and the PLAA was an early adopter of unmanned aircraft as intelligence, surveillance, and reconnaissance platforms in support of its artillery—particularly its long-range rocket artillery. CA-BDE artillery battalions include fire-finding radars, battlefield surveillance radars, long-range electro-optical and infrared sensors, and sound-ranging equipment, in addition to well-equipped forward observers, both mounted and dismounted. The PLAA devotes significant training resources to the battlefield surveillance and targeting mission. Observers of all types are likely well trained, and integration between sensors and shooters is likely agile, redundant, and reliable.

## FIRE SUPPORT EQUIPMENT OVERVIEW

B-5. The PLAA employs a huge variety of guns, mortars, and rockets. China manufactures a variety of these systems for export. This makes assessing PLAA equipment challenging. Much like with armored vehicles, units of the same type may possess systems with very different capabilities. One notable trend is how the PLAA prioritized the modernization of its artillery systems. Self-propelled guns (SPGs) and MRLs received enormous investments and have largely completely modernized systems, while many towed and mortar systems can be dated back to at least the 1960s.

> *Note.* The equipment overview provided in this appendix is not intended to be exhaustive, but rather to introduce the most widely employed variants of major equipment types used by the PLA.

## MORTARS

B-6. The Type 87 mortar is the standard light mortar employed throughout the PLAA, reserves, and militia units. A simple design similar to U.S. and Soviet or Russian light mortars, it fires a variety of ammunition, including high explosives (HE), smoke, and illumination, out to roughly 4,000 meters. These mortars are the basic indirect firepower of the light combined arms battalion (CA-BN), with each battalion employing a battery's worth of mortars (12 to 16 tubes) in the firepower company. The Type 87 is man-portable, though it is rather heavy by 81 or 82-mm standards.

B-7. The W-99 mortar looks like a hybrid between a mortar and a light howitzer. A towed system, it is referred to by the PLAA as a rapid-fire mortar, and it is capable of an impressive rate of fire, far surpassing that of man-packed 81 or 82-mm mortar types. It does, however, require significantly more support, including a prime mover and more ammunition. It is not man-portable, but it represents a significant increase in the firepower available to the infantry battalion. This capability is not replicated in U.S. formations.

## TOWED HOWITZERS

B-8. The Type 54 howitzer is a simple 122-mm pack gun derived from the World War II-era Soviet M-30. Despite its advanced age, the M-30/Type 54 is still widely fielded and is almost certainly the most numerous light howitzer design in the world today. Chinese designs were periodically upgraded over the years, becoming lighter, increasing reliability and rate of fire, and simplifying march-order and emplacement procedures. While old, the Type 54's performance is roughly in line with modern 105-mm guns, though it is somewhat heavier than most contemporary light howitzers. It provides the bulk of the artillery firepower for the light CA-BDE, and it still equips many of the towed howitzer battalions of the artillery brigades. The most common evolved versions of the Type 54 are the Type 83 and Type 86, both redevelopments from the 1980s that feature superior chassis technology and increased rates of fire. The Type 54 family is a versatile system. It can provide both indirect or direct fire, and PLAA forces train to employ these systems as direct-fire antitank guns in addition to their role as artillery.

B-9. The Type 66 howitzer is another Soviet design, based around the D-20 152-mm howitzer. This system dates from the 1960s, and it still comprises most of the PLAA's heavy tube artillery capability. The Type 66 is a large and heavy gun with middling-to-poor performance by modern standards, but when employed competently and in mass, it can still deliver decisive firepower. The system is simple and robust, able to operate in austere environments with minimal maintenance support. Its rate of fire is similar to that of more modern guns, but its range and accuracy are significantly worse when firing conventional ammunition. Rocket-assisted ammunition and precision-guided rounds offset these limitations somewhat, but their availability is limited. The Type 66 equips the heavy towed batteries of PLAA artillery brigades, and it is used to provide heavy fire support for maneuver operations.

## SELF-PROPELLED HOWITZERS

B-10. The PLZ-89 is the PLAA's most widely used SPG. It mates the gun from the ubiquitous Type 54/83/86 122-mm towed howitzers with the chassis of the Type 63 light tank, creating a simple, lightweight, mobile self-propelled artillery system. The performance and protection of the PLZ-89 are both unremarkable, but the system is robust and easy to employ. The PLAA employs around 700 PLZ-89s throughout its formations; nearly all are in the artillery battalions of medium or heavy CA-BDEs. These systems provide close-range fire support for CA-BDE maneuver actions, typically rolling just behind the maneuver battalions. Though outranged and outgunned by more-modern 155-mm SPGs, the PLZ-89's mobility and simplicity make it indispensable to the PLAA. The SH-3 is a recent modernized variant, offering improved fire control and mobility.

B-11. The PLZ-07 is the PLAA's newest 122-mm SPG. It employs a modernized version of the PLZ-89's gun, itself derived from the obsolete Soviet D-30, though its performance is improved with modernized fire control and advanced ammunition. The primary advance of the PLZ-07 is the use of the modern ZBD-04 armored personnel carrier (APC) as its chassis, giving it improved cross-country and amphibious performance, greater range, and improved reliability versus older SPGs. The PLZ-07 will likely be produced in large numbers, eventually replacing all of the older PLZ-89 subtypes throughout the PLAA. Like the older system, the PLZ-07 is outgunned by 155-mm systems, but its mobility and simplicity make it very well adapted for closely supporting maneuver formations.

B-12. The PLZ-83 was the PLAA's first heavy tracked SPG, mating the gun from the Type 66 howitzer with a generic Soviet heavy armored fighting vehicle (AFV) chassis. Though the 152-mm gun provided enhanced firepower compared to the widely fielded 122-mm models, the PLAA was dissatisfied with the PLZ-83's mobility and reliability. The ballistic performance of the gun was also poor. Relatively small numbers were fielded, with the PLAA initially equipping division artillery units with them, and today equipping individual batteries within artillery brigades with them. These systems are still in service, though they are set to be replaced by more-modern systems.

B-13. The PLZ-05 is the PLAA's new heavy SPG, mating a new 155-mm gun with a specially developed light AFV chassis. The PLZ-05 can employ a wide variety of munitions and enjoys the impressive range and accuracy of a fully modernized 155-mm gun system. In addition, the PLZ-05 employs an autoloader, increasing its sustained rate of fire to 10 rounds per minute, with a 3-rounds-in-15-seconds burst capability. The PLZ-05 is comparable to other advanced SPGs in most respects, and it represents a major increase in indirect-fire capability for the PLAA. It is likely that all artillery brigades will eventually employ a battalion's worth of PLZ-05s as the older PLZ-83s are phased out. The PLZ-05 is a very large and heavy system with substantial support requirements, but its cross-country mobility remains excellent.

## MULTIPLE ROCKET LAUNCHERS

B-14. The PHL-81 is the PLAA's most widely distributed 122-mm MRL. A simple truck-mounted design derived from the Soviet BM-21, the PHL-81 provides both CA-BDE artillery battalions and artillery brigades with long-range destructive firepower. Though the PHL-81 is derived from a 1960s-era design, it has been periodically upgraded with digital fire control and a wide variety of munitions, including HE, antitank submunitions, fuel-air, smoke, and mine scattering. More-specialized types of ammunition, including radio- frequency jamming, incendiary, and chemical, have been produced in the past or in other nations, and they may be available to PLAA forces. Precision-guided 122-mm rounds may also be available, though their cost and complexity limit their proliferation. Nominal range for the modernized 122-mm rounds is around 50 kilometers (kms). The PHZ-89 is essentially a tracked version of the PHL-81, developed to operate in support of PLAA mechanized and armored forces. It fires the same suite of 122-mm rocket rounds while improving protection and mobility at the expense of weight and cost. These systems feature a rapid-reload capability, enabling a second rocket pod to be fired within minutes and essentially doubling a battery's firepower against a time-sensitive target. Both the PHL-81 and the PHZ-89 are set to be replaced with domestically produced systems in the relatively near future. The PHL-11 is the new wheeled variant, and the PHZ-11 as the tracked.

B-15. The PHL-03 is the PLAA's version of the Soviet or Russian BM-30 Smerch, a widely proliferated heavy (300-mm) MRL. PHL-03 batteries provide long-range striking power for the PLAA's artillery brigades. Advancements in the 300-mm class of rockets make these among the most powerful and versatile artillery pieces fielded today. The PLAA operates a wide variety of 300-mm rocket ammunition types, including HE, antitank submunitions, fuel-air, incendiary, radio-frequency jamming, mine scattering, near-precision, and precision. New long-range variants push engagement ranges out more than 150 kms while maintaining reasonable accuracy. Reload of the PHL-03 is relatively slow. Each tube must be reloaded separately, rather than exchanging a pod. The PLAA likely views its 300-mm rocket class as the backbone of its operational fires capability of the future, bridging the gap between its gun and light rocket artillery and its ballistic missile force, and it is the primary offset for the lack of a meaningful close air support capability.

## BALLISTIC MISSILES

B-16. The <u>DF-11</u> is the People's Liberation Army Rocket Force's (PLARF's) most widely employed short-range ballistic missile (SRBM). Originally designed to provide operational-level area fires out to 300 kms to theater commanders, subsequent versions have expanded the range out past 700 kms. Accuracy has also increased, reducing circular error probable to only 30 meters, giving theater commanders a long-range precision strike capability. The DF-11 can employ both conventional and nuclear warheads; conventional warheads may be either HE submunition or fuel-air explosive. The PLARF is capable of structured attacks with its DF-11 fleet, and the solid-fuel rocket and mobile transporter-erector-launchers enable rapid launch and reload operations.

B-17. The <u>DF-15/16</u> is a newer family of systems that provides the bulk of the PLARF's operational-level fire support capability. These systems range between 600 kms and 1,000 kms, and they employ advanced antiballistic missile countermeasures such as terminal maneuvers and decoys. Early variants were not accurate enough for precision strikes, but modernized variants enjoy a circular error probable of 30 meters or less. These missiles can employ nuclear or conventional warheads, and have a significantly larger payload (over 600 kilograms) than do most SRBMs.

B-18. The <u>B-611</u> is a new SRBM system that likely seeks to replicate the capabilities of the Soviet SS-26 Iskander. Its capabilities include a 500-km range, advanced penetration aids, good mobility, rapid reload, and precision targeting. The B-611 is likely intended to complement the 300-mm rocket class with an enhanced penetration weapon, enabling the targeting and destruction of antiballistic missile systems prior to more general targeting.

*Note.* This section only discusses PLARF ballistic missiles that may be employed in support of tactical formations; it does not include long-range or intercontinental ballistic missiles (ICBMs).

## FIRE SUPPORT CAPABILITIES AND LIMITATIONS

B-19. The PLAA relies heavily on fire support to offset capability shortcomings in its maneuver forces and air power. Artillery of all types—mortars, tubes, and rockets—is deployed in significant density across every type of PLAA formation. Fire support is integrated effectively with maneuver operations. In particular, SPGs train to maneuver closely with armored and mechanized forces. Targeting can be decentralized, with lower echelon commanders integrating artillery spotters, unmanned aircraft (UA), and other surveillance assets. Massed fires are still considered the primary objective of artillery forces, and artillery commanders will generally seek to mass fires on the most valuable targets within range of their systems.

B-20. PLAA towed systems are much older and heavier than their Western counterparts, but they still have adequate tactical mobility to accomplish their fire support tasks. The PLAA relies on density and mass to accomplish these tasks, and logistics support for artillery is likely one of its primary limitations. The PLAA has a lower density of logistics support than do Western militaries—particularly the U.S. Army—and so sustaining ammunition requirements for operational artillery units may prove a significant challenge. PLAA SPGs are considerably lighter than their Western counterparts, and they are both outranged and outgunned, but they have fantastic cross-country mobility and easier sustainment requirements. The ability of these lightweight SPGs to maneuver closely with armored units is a key element of PLAA tactical operations.

B-21. PLAA rocket artillery is versatile and effective. It mixes a wide variety of different effects through the use of different munitions, integrated on simple, robust, mobile chassis. Lighter systems lack the destructive power of heavier Western MRLs, but their simplicity and versatility make them indispensable. Heavy MRLs comprise a huge part of the artillery brigade's firepower. These systems are carefully integrated with forward observers and other surveillance assets to make the most of their destructive potential.

B-22. While the PLAA fields a variety of precision and near-precision munitions for all of its fire support systems, the cost and complexity of these munitions limit their usage to the highest-priority missions. Precision munitions use a variety of guidance methods, including laser targeting, inertial, satellite, imagery, and radar. High-end ballistic missiles employ multiple methods to enhance accuracy at extreme ranges. The PLAA process for mensurating a target is not well known, but is likely less precise than U.S. processes. The decision-making process supporting targeting, however, is likely well-rehearsed and meticulous—in part due to reactions from the bombing of the Chinese embassy in Belgrade by American aircraft in 1999.

B-23. Like their Russian counterparts, PLAA fire support elements are very concerned with the counterfire threat, and significant resources are given over to counterfire operations. PLAA artillery units generally have good tactical mobility and practice the simple "shoot-and-scoot" methodology to avoid counterfire, while carefully concealing their positions from aerial attack and observation. The PLAA employs a variety of counterfire systems including sounding, radars, and visual means, and counterfire missions are highly prioritized.

This page intentionally left blank.

# Appendix C
# Air Defense Capabilities

## AIR DEFENSE FUNCTIONAL OVERVIEW

C-1. Ground-based air defenses are ubiquitous throughout the People's Liberation Army Army (PLAA), attempting to offset shortfalls in joint airpower. Experiences from Allied air power during the Korean War— and careful observation of the effects of joint airpower during Operation Desert Storm—profoundly influenced the People's Liberation Army's (PLA's) belief in air defense as a critical battlefield capability. Every maneuver formation larger than a company fields an organic air defense capability, and larger organizations employ a mixture of air defense systems to create a combined arms effect. Mass and depth are the operative words for PLAA air defense. Most of its short-range systems are relatively primitive and lack capability against modern air threats. They are, however, deployed very densely and broadly across the battlefield, creating a deterrent effect against low-altitude air threats. Medium- and long-range systems are significantly more lethal, capable of standing off against advanced air threats at any altitude.

C-2. Man-portable air defense systems (MANPADS) and heavy machine guns are fielded in the battalion's firepower company, and a mixture of self-propelled antiaircraft guns (SPAAGs), towed guns, and short-range infrared surface-to-air missile (SAM) systems, some medium-range radar systems, and a sensor and command component comprise the brigade's air defense battalion. Group army air defense brigades mix gun systems, short-range infrared missile systems, and medium-range radar missile systems. Air defense brigades also add an electronic air defense battalion that employs a variety of electronic warfare systems focused on defeating or neutralizing the enemy air threat. PLAA systems may or may not be fully digitally integrated into a wider integrated air defense system (IADS). Army units have the capability for digital integration all the way down to individual gun or missile teams, but it is not clear if this capability is fielded. Procedural controls such as airspace control measures and weapon control statuses are likely the dominant methods of fire control for most PLAA air defense units, excepting the medium-range SAM units.

C-3. Longer-range SAM systems are operated by the People's Liberation Army Air Force (PLAAF) as part of a nationwide IADS. PLAAF long-range SAM battalions are fully digitally integrated and likely centrally controlled by theater commands in cooperation with fixed-wing defensive counterair operations. Medium- range SAM units are also operated by the PLAAF, used either to defend long-range systems from oblique or pop-up attacks or to bridge gaps in SAM coverage. The PLAAF envisions its long-range SAM systems as wide area denial weapons, able to deny the use of airspace to enemy aircraft and create air superiority, or at least air parity, from the ground. These systems employ advanced counter-precision-guided munitions and electronic counter-countermeasures along with a variety of interceptor types and advanced, networked sensors.

## AIR DEFENSE EQUIPMENT OVERVIEW

C-4. PLAA air defense capabilities are primarily comprised of guns—towed, self-propelled, and hybrid— and missile systems that range from man portable to long range.

---

*Note.* The equipment overview provided in this appendix is not intended to be exhaustive, but rather to introduce the most widely employed variants of major equipment types used by the PLA.

---

## TOWED GUNS

C-5. The <u>PG-87</u> is the most widespread towed antiaircraft gun in the PLAA's inventory, with several thousand units equipping light combined arms brigades (CA-BDEs), reserve units, and militia units. The PG-87 is an up-gunned variant of the widely proliferated Soviet ZU-23, first fielded in 1960. PLAA variants employ two barrels firing 25-millimeter (mm) ammunition—in contrast to the Soviet 23 mm— giving Chinese weapons increased range and lethality. The PG-87 was further upgraded with enhanced targeting (including an infrared sight and fire control radar), improved training and direction mechanisms, and larger ammunition magazines. The PG-87 can fire a variety of ammunition types, including high explosive, armor piercing, and tracer. The PG-87's range is limited to two kilometers (kms), but it is lethal against aerial targets of all types that enter its engagement envelope. The PLAA also employs its PG-87 guns in direct-fire mode against ground targets. Its 25-mm ammunition has similar performance to infantry fighting vehicle (IFV) autocannons in this role.

C-6. The <u>PG-99</u> is a newer towed twin 35-mm system likely derived from the Oerlikon GDF family of guns. The PG-99 makes use of a variety of different search and tracking sensors, including an electro-optical visual sensor, a laser range finder, and digital integration with supporting radars. Traverse and elevation are rapid and mechanically assisted, and the gun has excellent mobility. Its emplacement time is less than two minutes. The PG-99 can fire high explosive, incendiary, and armor piercing rounds past four kms, and it can also engage ground targets, including light armored vehicles. The PG-99's gun system is also the primary weapon for many of the PLAA's SPAAGs and hybrid systems, simplifying sustainment and crew training requirements.

## SELF-PROPELLED GUNS AND HYBRID SYSTEMS

C-7. The <u>PGZ-04/04A</u> marries the ubiquitous ZBD-04 armored personnel carrier (APC) chassis with four 25-mm autocannons taken from the PG-87 towed gun and four QW-2 short-range infrared missiles, creating a mobile, protected short-range air defense (SHORAD) platform. The PGZ-04 enjoys the same protection as its APC and IFV counterparts, and it likely has comparable mobility. Its sensor suite includes the same electro-optical, infrared, and laser rangefinder as the PLAA's towed guns and it adds a short-range, low-altitude search radar. The 25-mm guns have a range of around two kms, while the QW-2 missiles can reach out to six kms. Firing doctrine likely instructs crews to employ guns against low and slow targets such as unmanned aircraft, reserving the missiles for standing off more-dangerous high-performance threats such as helicopters. Crews train to employ the 25-mm guns against ground targets as well. The combination of the advanced electro-optical, infrared sensor and the four autocannons likely make the PGZ-04 a particularly effective ground support system.

C-8. The <u>PGZ-07/09</u> is a heavy SPAAG built on the PLZ-05 chassis. It mates the 35-mm gun from the PG-99 towed system with a short-range radar system. The PGZ-09 may also employ short-range infrared missiles (likely the same QW-2 as the PGZ-04), though this is unconfirmed. The 35-mm guns employ a variety of ammunition types out past four kms, and they can also engage ground targets: the firepower of the PGZ-09's guns in a direct-fire role surpasses nearly every existing IFV. The PGZ-07s complement the PGZ-04s in the heavy CA-BDE's air defense battalion, and they are also fielded in the air defense brigade.

## MAN-PORTABLE MISSILE SYSTEMS

C-9. The <u>HN-5</u> is China's first MANPADS, developed from the early variants of the Soviet 9K32 Strela-2. China was late in developing MANPADS as compared to the U.S. and the Soviets, and by the time the HN-5 reached operational status it was already obsolete. Upgraded variants were produced throughout the 1980s and 1990s, including an all-aspect version, though these too are now largely ineffective against modern countermeasures. The HN-5 claims a range of four kms, but this distance is significantly limited by aspect angle. HN-5 variants are employed as man-pack weapons and mounted on tactical vehicles of all types—particularly light wheeled vehicles—mated with a digital electro-optical fire control sensors. The remaining HN-5/5A/5B rounds are employed by reserve and militia units, and they may still effectively engage larger unmanned platforms or older military aircraft that lack modernized MANPADS countermeasures.

C-10. The <u>FN-6</u> is a reverse-engineered copy of the European Mistral system. Though technically still a MANPADS, the FN-6 is a heavier and more-powerful missile than most MANPADS, and it is best suited for use on vehicle-mounted platforms or from stabilized ground positions. The PLAA employs FN-6 variants on a wide variety of vehicles, most notably the <u>FN-6A</u>. The FN-6A combines the infrared missile with a heavy machine gun on a light tactical vehicle chassis, mimicking the U.S. Army's AN/TWQ-1 Avenger SHORAD platform. It is likely that most future SHORAD vehicles will employ the FN-6 and evolved variants—such as the <u>FN-16</u>—as either a dedicated SHORAD system or a bolt-on option, as required. The FN-6 is an all- aspect weapon with a range of more than six kms.

C-11. The <u>QW-1/2</u> are the PLAA's newest and most advanced MANPADSs. They are likely reverse-engineered or rough copies of the Russian 9K310 Igla-1 system. Each is a lightweight, shoulder-fired system with an advanced infrared seeker and continuously upgraded counter-countermeasures. The newer QW-2 has been upgraded to increase its performance against very low and fast-moving targets, potentially giving it a counter-cruise missile capability. The QW series is also mounted on a vast array of vehicles, ranging from light wheeled systems to heavy armored systems. Most of these systems employ only visual and infrared sensors, though some integrate short-range radars. QW-series missiles have a range of roughly six kms.

C-12. Chinese MANPADS systems are enabled by an advanced command and control system called SmartHunter. This system includes an advanced short-range radar, identification of friend and foe capability, and network backbone to support decentralized operations by MANPADS teams. SmartHunter is vehicle-mounted and tactically mobile, able to provide a high-quality air defense sensor and fire control support in advanced or austere areas.

## SHORT-RANGE MISSILE SYSTEMS

C-13. The <u>HQ-6D</u> is a short-range system that combines a semi-active radar-homing interceptor—derived from the U.S. AIM-7 Sparrow—and a truck-mounted transporter-erector-launcher (TEL). An HQ-6D battery includes a medium-range surveillance radar, tracking radars, and up to eight TELs, each with four ready missiles. The HQ-6D is less lethal and less survivable than more-modern active radar-homing systems, but it is still used for point defense missions against low-flying aircraft and cruise missiles. Evolved versions can be centrally controlled at the battalion echelon, and they can be integrated into a theater IADS. The HQ-6D's interceptor has a slant range of roughly 18 kms. The HQ-6D is the primary short-range system employed by the PLAAF.

C-14. The <u>HQ-7</u> is a widely proliferated short-range infrared system based on the French Crotale. Though relatively old by SAM standards—it was first fielded in the early 1980s—the HQ-7's simplicity, mobility, and lethality have contributed to a long and ongoing service life in the PLAA. HQ-7 interceptors are deployed on mobile trucks, at fixed sites, and on ships. A battery consists of a short-range search radar and three launchers, each with either four or eight ready missiles. Launchers can be cued by the radar or engage targets visually. The interceptor's slant range is roughly eight kms against a fast-moving target and up to 15 kms against a slower target. The HQ-7 is typically the primary missile system of the medium CA-BDE, with one or two battery's worth of equipment organic to that organization. Other batteries are employed in static defense roles of critical assets. The HQ-7 is employed by both the PLAA as a mobile SAM platform and by the PLAAF as a lower-tier system.

## MEDIUM-RANGE MISSILE SYSTEMS

C-15. The <u>HQ-16/16A</u> is a highly modified variant of the Russian 9K37 Buk. The HQ-16 combines a modernized semi-active radar-homing interceptor with a mobile tactical six wheel drive truck. Its interceptor was originally procured as a naval system, then adapted to land operations once its potential was recognized. A battery consists of a command post, two multi-mode (surveillance/tracking/illumination) radars, and four to six TELs, each with six ready missiles. The HQ-16 found a useful home within the PLAA. Its mobility makes it difficult to target when part of joint suppression of enemy air defenses, and its range/lethality occupy the space between the HQ-6/7 and the HQ-9. The HQ-16 is used primarily as a point defense system for critical assets, especially those vulnerable to attack by cruise missiles. Its interceptors are relatively simple and inexpensive, making it an attractive option for wide proliferation. The original interceptor's slant range was around 40 kms. This has been extended past 70 kms in the newest HQ-16B/C variants. The PLAA operates the HQ-16, but it likely integrates it fully into theater IADS along with the long-range PLAAF SAM systems.

C-16. The HQ-17 is a reverse-engineered copy of the Russian 9K331 Tor-M1 SAM system. The HQ-17 marries a simple command-guided interceptor with a robust tracked chassis and a powerful electronically scanned radar array, creating a mobile and protected medium-range SAM system. HQ-17s are primarily housed within heavy CA-BDEs, providing a uniquely lethal and long-range air defense capability within the CA-BDE's air defense battalion. HQ-17 batteries consist of a medium-range radar and four launchers, each with eight ready missiles. Though it uses an older command guidance system, the HQ-17's accuracy and lethality are thought to be excellent. Some analysts claim that the HQ-17 is capable of challenging targets such as cruise missiles and precision munitions. Its slant range is quoted as anywhere between 12 kms and 20 kms; the detection range of the radar is past 30 kms.

## LONG-RANGE MISSILE SYSTEMS

C-17. The HQ-2/A/B is an evolved version of the aging Soviet S-75/SA-2, one of the earliest effective SAMs. Modernized HQ-2 batteries employ a multi-mission radar instead of the numerous single-purpose radars of earlier versions, but the command-guided interceptor remains largely unchanged. The slant range of PLAAF HQ- 2 interceptors is roughly 45 kms. The HQ-2 lacks the maneuverability and precision of more-modern systems, but its speed and range still pose a threat to larger, less-maneuverable aircraft. It is likely that the large numbers of HQ-2 systems still in use are envisioned as deception or decoy platforms, forcing the enemy to alter plans or expend expensive and rare joint suppression of enemy air defense resources to defeat or neutralize obsolete, inexpensive equipment. Each HQ-2 battery includes a multi-mission radar and up to six launchers, each with one ready missile.

C-18. The HQ-12/22 was originally intended as an incremental improvement to the HQ-2, but due to a lengthy development timeline it entered service at about the same time as the next generation of SAM systems. It employs a semi-active radar-homing interceptor and a multi-mission, phased-array sensor. It is truck mounted and significantly more mobile than the HQ-2—though not sufficiently so to support maneuver forces. Initial variants had a slant range of around 70 kms. The newer HQ-22 pushes this out to around 170 kms. An HQ 12/22 battery consists of a multi-mission radar and four launchers, each with two ready missiles. While the HQ 12/22 does not have performance comparable to that of the HQ-9 and newer SAM systems, it is far less expensive, and it is used to reinforce more-advanced SAM deployments with adequate, less-expensive interceptors and radars.

C-19. The HQ-9/HQ-15/HQ-18 is a family of long-range SAM systems based around the Russian S-300. Beginning with the HQ-9, these systems represent one of the largest and highest-profile military programs in Chinese history, helping to form the backbone of the PLA's anti-access and area denial strategy. The S-300 was built using a system-of-systems concept, enabling a battery or battalion to employ a mixture of radars and interceptors to meet operational needs. The PLA began development of the HQ-9—a reverse-engineered S-300—after carefully observing the devastating effects of allied air power during OPERATION DESERT STORM. The HQ-9 was specifically designed to deny large areas of airspace to fourth-generation Western aircraft through a combination of advanced sensors; large, long-range interceptors; and modernized electronic counter-countermeasures. While initial variants lacked the kinematic performance of contemporary Russian and U.S. SAM systems, periodic upgrades dramatically improved both the speed and maneuverability of interceptors. The HQ-15 was the first major modernization, adding an antiballistic missile capability and extending both detection and engagement ranges past 200 kms. The HQ-18 was the second major upgrade, adding a new pair of high-performance interceptors along with a larger, more-powerful radar. The HQ-18's interceptor ranges are shorter, but its performance against the most challenging targets—ballistic and cruise missiles, other precision-guided munitions, and maneuvering aircraft—is likely superior to the HQ-15. An HQ-9/15/18 battery typically consists of a long-range search radar, a powerful tracking radar, and up to eight TELs, each with four ready interceptors. Interceptors and radars can be mixed to a certain extent, either within a battalion or even within a battery. China recently purchased several S-400 systems from Russia. If the country follows its previous practices, these systems will be used to develop reverse-engineered upgrades to existing systems. All HQ-9/15/18 systems are integrated into theater IADS.

# AIR DEFENSE CAPABILITIES AND LIMITATIONS

C-20. Chinese air defense capabilities are comprehensive and numerous, mixing mobile, short-range systems under the control of the PLAA with powerful, modernized long-range systems operated by the PLAAF. Maneuver units will typically not operate without at least local SHORAD systems in support. MANPADS are widely distributed throughout PLAA formations of all types, and they are also employed by militia and reserve units throughout rear areas. This creates a tiered and layered air defense capability, using the combined arms effect of multiple systems to offset the vulnerabilities of one system with the strengths of another.

C-21. Long-range and many medium-range systems operate as part of the theater IADS. This integration amplifies the effectiveness of individual batteries by enabling them to participate in data sharing and centralized fire control. It is not known how effectively hardened the Chinese IADS communications network is, but it is likely among the highest-priority network systems in the PLA. Lower echelon units may also have the capability to participate in a wider air defense network construct, but they are not dependent on it. Battalions and batteries operate in a decentralized fashion, employing localized networks that integrate sensors and shooters at tactical echelons. This enables even the smallest units, such as a MANPADS team, to receive data and fire control support from their higher echelon headquarters.

C-22. The PLAA's continued interest in gun-based systems is unique among the world's advanced militaries: no other modern army employs such a density of gun systems. While guns have significant limitations in range and accuracy as compared to missiles, their simplicity, versatility, and capability against low-altitude air threats validate the continued PLAA investment. Indeed, several other militaries— the U.S. among them— are revisiting heavy antiaircraft guns, decades after abandoning them, as offsets to unmanned aircraft and attack helicopters.

This page intentionally left blank.

# Appendix D

# Aviation Capabilities

## AVIATION FUNCTIONAL OVERVIEW

D-1. People's Liberation Army Army (PLAA) aviation capabilities consist of a mixture of attack, multirole (medium), and transport helicopters, plus a number of different unmanned aircraft systems (UASs). Army aviation was a low priority of the PLAA for most of its history. This is reflected in the very low density of Chinese rotary-wing systems compared to other first-order militaries. As a basis for comparison, the U.S. Army's density of helicopters per capita is over eight times that of the PLAA. Nonetheless, the PLAA has invested significantly in developing its rotary-wing capabilities in recent years, moving from a force employing only light multirole helicopters in limited missions to fully equipped army aviation brigades (AABs) that field a mixture of new advanced attack, reconnaissance, multirole, and transport helicopters. AABs are likely to continue expansion as the PLAA modernizes.

## AVIATION EQUIPMENT OVERVIEW

D-2. PLAA aviation capabilities consist of attack, multirole, transport, and light helicopters and unmanned aircraft (UAs). These capabilities are described in paragraphs D-3 through D-10.

---

*Note.* The equipment overview provided in this appendix is not intended to be exhaustive, but rather to introduce the most widely employed variants of major equipment types used by the People's Liberation Army (PLA).

---

### ATTACK HELICOPTERS

D-3. The Z-10 is China's first modern attack helicopter. Pursued after the PLA studied the effectiveness of the U.S. Apache during Operation Desert Storm, the Z-10 is the result of collaboration between the Russian Kamov design bureau and the PLA. It strongly resembles other modern attack helicopters both physically and in capabilities, and its specifications are competitive with any contemporary design. The Z-10 has a crew of two and employs a chin-mounted chain gun—likely a copy of the U.S. M242 25-millimeter (mm) system—along with a versatile mixture of missiles and rockets. Z-10 missiles include both precision air-to-air and air-to-surface weapons. The Z-10 can carry up to 16 missiles of various types, four rocket pods, or some combination thereof, as required. The Z-10 also employs a modern sensor suite, including fire-control radars, infrared optics, and an optional targeting pod with an electro-optical targeting system. The Z-10 has an integrated electronic warfare capability that can be augmented by an add-on pod.

D-4. The Z-19 is a lightweight attack and reconnaissance helicopter that was developed in concert with the Z-10 and is derived from the Z-9 utility helicopter. It also employs a two-person crew, though it is substantially smaller and lighter than the Z-10. Unlike virtually every other attack helicopter, the Z-19 does not use a machine gun or cannon; its offensive capability comes either from eight missiles (air-to-air or air-to-surface) or two rocket pods. Its sensor suite is similar to that of the WZ-10, with a radar system and infrared targeting. The Z-19 features an integrated targeting pod under the nose, enhancing its targeting and reconnaissance capabilities at the expense of a gun system. Z-19s are fielded in the reconnaissance battalions of AABs; their primary tactical role is reconnaissance and targeting in support of heavier attack platforms. They may also be used more extensively in the rotary-wing air-to-air role; their small size, smaller infrared signature, and increased agility makes them well-suited to the low-level counter-air mission.

## MULTIROLE HELICOPTERS

D-5.  The <u>Mi-17</u> is a long-serving Russian medium-lift and -utility design still in limited use in the PLAA. An evolved version of the Mi-8, the Mi-17 is slightly larger and more powerful than the U.S. UH-60 Black Hawk. It is a simple utility design able to carry or sling a useful payload or up to two squads' worth of troops. Winglet-equipped variants can carry a substantial offensive payload, including rocket and gun pods, though they lack the sophisticated targeting capabilities and protection schemes of dedicated attack helicopters. The PLAA uses several different Mi-17 variants, though their performance and roles vary little. Mi- 17s serve mostly in the utility helicopter battalions of AABs.

D-6.  The <u>Z-9</u> is a light multirole helicopter derived from a European design. It is versatile and lightweight, able to serve in light transport, attack, reconnaissance, search-and-rescue, or maritime roles. In attack roles, it employs a mixture of light cannons and up to four missiles, both air-to-air and air-to-surface. The Z-9 may be considered a compromise platform. It was fielded as one of China's first modern military helicopters. It lacks the endurance and payload of a transport helicopter and the protection and sensors of an attack helicopter. Nonetheless, Z-9s are fielded widely throughout the PLAA, the People's Liberation Army Air Force (PLAAF) and People's Liberation Army Navy (PLAN). Z9s will see widespread use in multiple roles for the foreseeable future.

## TRANSPORT HELICOPTERS

D-7.  The <u>Z-8</u> is a modernized version of a 1960s-era French heavy-lift helicopter, the SA-321. China reverse-engineered and modernized the French design, making it the standard heavy-lift platform for both the PLAA and the PLAN. The Z-8 is both fast and long-ranged, and it can carry a substantial payload. The basic design is very old; however, and it is both inefficient and maintenance-heavy compared to more modern designs. Due to its size and poor agility, it is also very vulnerable to ground fire, and it is likely relegated to transport roles in deeper operational areas.

D-8.  The <u>Z-20</u> is a new design that bears a very strong resemblance to the UH-60 Black Hawk. It is likely reverse-engineered from copies of the S-70, the Black Hawk's civilian counterpart. The Z-20 is a new design, accepted into service in 2018, and few examples are operational. It is very likely, however, that the Z-20 will become the PLAA's new standard medium-lift and multirole platform, replacing both the Z-8 and the Mi-17 in these roles. Navalized versions are also in development. Unsurprisingly, the Z-20's specifications mirror those of the UH-60, though uprated engines and a new five-bladed propeller may moderately enhance performance.

## LIGHT HELICOPTERS

D-9.  The PLAA operates a number of light helicopter systems, nearly all of which are either commercial purchases from European manufacturers or licensed or reverse-engineered copies of European designs. The most common is the <u>HC-120</u>. It is a simple and agile design that performs a number of light transport and liaison duties. These systems do not appear to be used in combat roles. They are, however, likely used by both People's Armed Police (PAP) and militia units as light air support for rear area and security operations, and civilian variants are widely fielded by Chinese police organizations of all types.

## UNMANNED SYSTEMS

D-10. China is a world leader in UAS development and production. This industrial capability is reflected in the widespread use of UASs throughout PLA formations of all types. The PLAA employs advanced medium-altitude UA as surveillance platforms at the theater and group army echelons, and they have likely weaponized some of these systems. Lightweight UA are employed at brigade and battalion level, and man-portable systems are employed by units as small as squads or patrols. The PLAA also employs weaponized UA in antiradiation and electronic warfare roles. The PLAN uses UASs for long-range surveillance in support of surface operations, and it may employ weaponized versions as antisubmarine platforms. The PLAAF employs a variety of sophisticated long-range UASs that provide surveillance and intelligence at the national and theater level and weaponized systems that conduct precision strike missions.

# AVIATION CAPABILITIES AND LIMITATIONS

D-11. PLAA aviation is relatively modern, and it fields a number of capable systems. Its primary limitation is the low density of systems and limited maintenance and sustainment resources, especially at lower tactical echelons. The PLAA has enthusiastically pursued an air assault capability and, though it is still developing, this investment will likely continue. It is unclear how effectively ground forces integrate with attack aviation; the relationship between the AAB and the combined arms brigade (CA-BDE) is likely theoretically similar to that between the U.S. Army's combat aviation brigades and its brigade combat teams but, considering the newness of PLAA attack aviation, this integration is likely immature.

This page intentionally left blank.

# Appendix E
# Engineer and Chemical Defense Capabilities

## ENGINEER AND CHEMICAL DEFENSE FUNCTIONAL OVERVIEW

E-1. The People's Liberation Army Army (PLAA) employs a robust combat engineer capability throughout its maneuver formations, underpinning the consistent focus on obstacles, entrenchments, fortifications, and mobility seen throughout its doctrine. There are four primary components to the PLAA combat engineer capability: mobility, countermobility, protection, and recovery. Mobility systems include those designed to repair and maintain roads, establish or repair bridges, and sweep enemy minefields. Countermobility systems construct obstacles and deploy mines, while protection systems build entrenchments and fortifications. Recovery systems recuperate damaged or immobilized vehicles.

E-2. Chemical defense capabilities are considered part of the same capability set as engineering, and they frequently occupy the same unit. The chemical defense mission is twofold: equipping and training the parent unit for operations in a chemical environment and decontaminating equipment and personnel that have suffered chemical attack.

E-3. In addition to traditional combat engineering, the PLAA and People's Armed Police (PAP) both have robust civil engineering capabilities. Civil engineering missions are conducted in support of both military operations and civil authorities. The Chinese government views these operations as having great social value to the Chinese population. PLAA engineering units have among the most-advanced heavy equipment and engineering expertise available in China. Their projects include buildings, highways, bridges, railroads, air and seaports, and pipelines. Nominally, the PLAA assumes that many of these are dual-use projects or of some military value. In addition, PLAA engineering units stand by to serve as rapid-reaction forces in the event of natural disasters or other such events. Engineering support is one of the flagship examples of Military-Civil Fusion.

## ENGINEER AND CHEMICAL DEFENSE EQUIPMENT OVERVIEW

E-4. Historically, PLAA engineering work was performed either with simple manpower or with simple modifications to tactical vehicles—such as attaching a plow or bulldozer to the front of a tank. While a lot of this "do-it-yourself" mentality still exists within modern formations, the PLAA now has a portfolio of modernized engineering vehicles and specially trained personnel. Each combined arms brigade (CA-BDE) employs a variety of engineering systems in the operational support battalion, which includes a mobility company, a protection company, and a mine warfare company. Heavier engineering capabilities are housed in the group army's engineer and chemical defense brigade.

> *Note.* The equipment overview provided in this appendix is not intended to be exhaustive, but rather to introduce the most widely employed variants of major equipment types used by the People's Liberation Army (PLA).

E-5. The GCZ-110 is one of several types of tracked and armored multipurpose engineer vehicles, all of which are simply tank chassis and engine with the turret removed and engineering equipment attached. These systems provide the heavy and medium CA-BDEs with obstacle breaching, fortifications, and recovery support. Light CA-BDEs have lighter engineer support, consisting of similar capabilities but on truck-mounted platforms.

E-6. Mine laying and clearing remain important capabilities for the PLAA. Specialized mine-clearing vehicles combine plows and magnetic detectors with short-range mine-clearing rocket artillery on armored, tracked platforms, most notably the GSL-130. Lightweight wheeled systems are also common, serving in both medium and light CA-BDEs. Mine laying is accomplished primarily through artillery-scatterable mines, although large, advanced antitank mines are also employed in large numbers.

E-7. The PLAA employs a wide variety and high density of bridging systems, both truck-mounted and tank chassis-mounted. Lightweight bridging systems can handle wheeled or light armored systems, while heavy bridging systems can handle main battle tanks (MBTs). Many bridging systems are modular, allowing multiple segments of bridge to be combined together in order to cross wider gaps. PLAA maneuver units conduct regular training with bridging units and consider river crossing to be an important skillset. Bridging systems are also regularly used in support of civilian populations.

E-8. As with most militaries, the PLAA's armored recovery vehicles nearly all consist of tank chassis and engines, a fully armored crew compartment, and simple, powerful recovery gear designed to quickly retrieve damaged or immobilized armored vehicles. As the weight of armored vehicles—particularly tanks—has risen, so too have recovery vehicles needed to become more powerful. The PLAA has traditionally developed a recovery vehicle based on the previous generation's MBT. The most current recovery vehicle, the Type 654, is thought to be the only vehicle capable of recovering the Type 99 MBT. Heavy CA-BDEs employ recovery vehicles in their operational support battalion. Engineer brigades also operate recovery vehicle companies in support of group army operations.

## ENGINEER AND CHEMICAL DEFENSE CAPABILITIES AND LIMITATIONS

E-9. The PLAA's engineers are further along in their modernization timeline than several other Chinese capability sets. Modernized equipment and evolved tactics have been in place for several years, and frontline engineer units have conducted extensive training exercises using all aspects of these updated capabilities. In particular, armored engineering units appear highly capable in conducting obstacle breaching, mine clearing, and fortification construction in support of armored maneuver. Mine-laying systems also appear mature and capable. The high density of rocket artillery systems throughout the PLAA simplifies the use of scatterable mines at the CA-BDE.

E-10. Civil engineering support capabilities are also well developed, and the relationship between army engineers and civilian authorities is viewed as very important and meaningful. Work on civilian projects provides useful training opportunities to military engineers while improving local infrastructure. These projects are typically given high priority and significant media coverage.

E-11. The most significant limitations that PLAA engineers face are tactical mobility and sustainment. A large proportion of new PLAA engineer vehicles are heavy tracked systems, and they are thus ill-suited for movement over long distances by themselves. The PLAA has relatively few heavy transports capable of moving heavy vehicles, and these will likely be reserved for tanks and infantry fighting vehicles (IFVs) in the event that an armored unit needs to conduct a rapid overland movement. Sustainment is likely another issue. These systems have significant fuel and maintenance requirements. The relatively low density of PLAA logistics and armament troops may struggle to sustain heavy armored forces over long distances or for an extended period of time. This issue is further compounded by maneuver forces likely having a higher priority for sustainment assets than engineer forces. The PLAA does enjoy, however, a robust strategic transportation network within Chinese borders, based mostly around China's enormous rail network. This network is capable of moving large bodies of troops and equipment in and around China rapidly and efficiently, but it has little reach outside Chinese territory.

# Network and Communications Capabilities

## NETWORK AND COMMUNICATIONS FUNCTIONAL OVERVIEW

F-1. The People's Liberation Army Army (PLAA) historically did not seek to integrate tactical echelons into a wider communications architecture; the division was the first echelon with a robust communications capability. Lower echelons relied on a combination of close proximity to headquarters and couriers to meet their communications needs, with battalions as the lowest echelon to employ modernized radios. The evolution of the PLAA into an informationized force required a complete overhaul of the network and communications backbone, along with a far more robust density of communications systems and specialists. These reforms are ongoing. As with most of the PLAA's modernizations, units are likely in various stages of development. The PLAA's desired end state is likely to have every squad or patrol equipped with a secure, reliable radio communications capability and to extend a secure, reliable network data capability down to the platoon. Air defense and artillery systems are to be part of a theater-wide sensor-to-shooter network. Electromagnetic attack and defense capabilities are linked to tactical commanders, and they support tactical-level operations.

## NETWORK AND COMMUNICATIONS EQUIPMENT OVERVIEW

F-2. PLAA radio sets largely mirror those employed by U.S. and other North Atlantic Treaty Organization (NATO) forces. This is unsurprising—most PLAA radios are derived from U.S. designs acquired during the 1970s and 1980s. Company and battalion radio networks are built around a powerful tactical command post radio, with a series of either man-packed or vehicle-mounted radios connecting to an encrypted network. Platoon leaders employ backpack or vehicle systems such as the TBR-121, which has many of the same capabilities as the most modern U.S. single-channel ground and airborne radio system (also known as SINCGARS). Encrypted voice communications and limited data transfer are possible over the air.

> **Note.** The equipment overview provided in this appendix is not intended to be exhaustive, but rather to introduce the most widely employed variants of major equipment types used by the People's Liberation Army (PLA).

F-3. Chinese artillery and gun radio networks are more mature than other Chinese communication systems. Forward-observer networks employ powerful two-way systems that integrate observers and guns over secure data and voice networks, based largely on the TBR-142, a component-based radio system of systems. These networks can be very organized, linking gun pairs with individual observers, or they can be centralized and controlled through a fire-direction section. Unmanned aircraft system (UAS) integration into forward-observer nets is likely performed at fire-direction sections as well, allowing an artillery battalion to integrate multiple sensor types into its network. Artillery brigade systems are more advanced and powerful than battalion systems, allowing greater data throughput and range in support of heavy rocket and howitzer batteries.

F-4. PLAA communications units now employ both satellite-based tactical communications and terrestrial data networks. Satellite communications are heavily limited by bandwidth, and they and are reserved for the highest-priority networks. China plans to launch numerous new communications satellites in the coming years, however, and this may expand the availability of satellite networks to lower echelons. The PLA has a unique relationship with Chinese civilian telecommunications networks. Many civilian networks were built with PLA requirements in mind. As a result, the PLA can employ domestic cellular and landline networks for both voice and data communications. This capability is limited strictly to domestic military operations. Any expeditionary operations must use secure tactical communications on separate networks.

F-5.   Integrated air defense system (IADS) components employ their own data network that integrates sensors and shooters from both the PLAA and the People's Liberation Army Air Force (PLAAF). While track-sharing and sensor fusion are likely not yet fielded capabilities, medium- and long-range platforms can likely share airspace pictures and voice networks, and fire-control centers are integrated with fixed-wing assets as part of defensive counterair operations. The static nature of most long-range surface-to-air missile systems enables most of these network components to be hardened and isolated, making them immune to jamming or intrusion. Lower-level air defense systems use a mixture of voice and limited data to direct fire. Larger systems may have an automatic cueing capability. Man-portable air defense systems (MANPADS) can receive fire-control orders from a fire direction section, and they may also have small data devices to enhance situational awareness.

# NETWORK AND COMMUNICATIONS CAPABILITIES AND LIMITATIONS

F-6.   The expansion of modern communications capabilities to lower echelons has been challenging. The increased quality of PLAA recruits has helped to smooth this transition, providing a cohort of tech-savvy conscripts to operate advanced communications equipment. The PLAA is still developing a modern radio network, however, and it will inevitably run into significant roadblocks as its forces are modernized. These roadblocks will likely include everything from dealing with frequency management issues in a contested electronic environment to a lack of batteries at lower tactical echelons.

F-7.   While terrestrial radios are a meaningful first step to an informationized force, the challenges of building an expeditionary force—and the consistently challenging terrain of China—highlight the PLA's shortfalls in satellite, long-range, and non-line-of-sight communications technologies. While the PLA has a joint data network similar to Link-16, it is still in its infancy, and it is likely many years away from being useful in combat. Integration between land, air, and naval assets is poor to nonexistent—a major limitation in achieving an integrated joint force. Higher echelon units, such as theater commands, have access to advanced, specialized equipment such as troposcatter radios, but these are not fielded at tactical echelons. Communications between tactical units over wide areas is a challenge due to limited resources.

F-8.   The PLAA places a very high priority on electromagnetic defense as a part of an information superiority campaign and carefully encrypts as many tactical communications as possible. Data networks are hardened against intrusion, and specialized cells at the brigade and group army echelons conduct ongoing electromagnetic and communications network protection missions.

# Appendix G

# Special Operations Forces Capabilities

## SPECIAL OPERATIONS FORCES FUNCTIONAL OVERVIEW

G-1. China's special operations forces (SOF) represent the best-trained and best-equipped light ground forces in every People's Liberation Army (PLA) service. The People's Liberation Army Army (PLAA), People's Liberation Army Navy (PLAN), People's Liberation Army Air Force (PLAAF), and the People's Armed Police (PAP) all operate units designed as SOF. Each group army fields an SOF brigade, with specialized training and equipment suited to its theater of operations. In addition, the PLAN employs at least one dedicated SOF brigade, the PLAAF employs an airborne army, and the PAP employs at least two highly specialized SOF groups.

G-2. PLAA SOF units have a fundamentally different task and purpose than U.S. military special forces, the United States Special Operations Command (USSOCOM). USSOCOM missions include direct action, strategic reconnaissance, foreign internal defense, unconventional warfare, counterterrorism, and civil affairs. PLAA SOF, conversely, focus on support of conventional military operations. Special reconnaissance is their most-important mission. Their other primary missions include sabotage, raids, deep targeting, and search and rescue. As such, PLAA SOF can be viewed as an elite light infantry capability, as opposed to a Western-style SOF capability. PLAA SOF brigades at the group army will be employed in much the same way as the U.S. Army employs its Ranger light infantry units. PAP SOF units, on the other hand, focus more on security, counterterrorism, and hostage rescue missions, and they are more comparable to elite U.S. domestic law enforcement units. It is likely that these two SOF missions overlap in combat situations, with PLAA SOF taking a more forward or deep role and PAP SOF a more rearward role.

## SPECIAL OPERATIONS FORCES CAPABILITIES AND LIMITATIONS

G-3. Chinese SOF brigades generally receive the best-quality equipment, the highest training priority, and the best recruits and officers. As a result, their training and readiness level is likely comparable to that of the U.S. Army's best light infantry units. Unlike most Western SOF units, however, they do not always recruit experienced soldiers from the active-duty force. Quality conscripts may be recruited directly into SOF training, and new officers can be selected directly for SOF assignments.

G-4. Chinese SOF units, particularly those in the PLAA, are carefully tailored to the regions in which they serve. Units in mountainous regions receive mountain training, units in jungle areas receive jungle training, units in maritime areas receive amphibious training, and so on. All PLAA SOF units focus on urban combat, particularly small-unit tactics in urban environments. All Chinese SOF units are airborne and air assault capable, though no Chinese SOF unit has the same access to air support as USSOCOM units. The PLAA and PLAAF lack the aviation capabilities to conduct the deep clandestine insertions necessary for strategic-level direct action or reconnaissance, but they can conduct air insertions of SOF in support of operational units.

G-5. There are no PLAA SOF units specially designed or equipped for long-distance expeditionary operations, but units do conduct joint training exercises with other national militaries. Outside of operations in and around Chinese border areas and territorial waters, PLAA SOF are generally domestically oriented. Antiterrorism and riot control are key missions for both PLAA and PAP SOF, and either of these units may assist in maintaining domestic stability in conjunction with police or other security personnel.

G-6. PLAA SOF do not conduct foreign internal defense or unconventional warfare, with the possible exception of areas in the immediate geographic area of China. It is possible that these missions will be adopted as the PLA develops its expeditionary capability. In general, it is likely that the Chinese government will prioritize the continued development of its SOF units, as these capabilities provide greater flexibility and deployability than heavy conventional forces, enabling greater force projection and providing a wider range of military and nonmilitary options to political leaders.

# Glossary

The glossary lists acronyms and a term with a joint definition. Acronyms appearing in ATP 7-100.3 that are not Army or joint are marked with an asterisk (*).The proponent publication is listed in parentheses after the definition.

## SECTION I – ACRONYMS AND ABBREVIATIONS

| | |
|---|---|
| *AAB | army aviation brigade |
| ADP | Army doctrine publication |
| *AFV | armored fighting vehicle |
| APC | armored personnel carrier |
| AR | Army regulation |
| ATGM | antitank guided missile |
| ATP | Army techniques publication |
| B.C. | [years] before Christ; equivalent to [years] B.C.E. ("before common era") |
| BCT | brigade combat team |
| CA-BN | combined arms battalion |
| *CA-BDE | combined arms brigade |
| CAS | close air support |
| CCG | China Coast Guard |
| *CMC | Central Military Commission |
| *CNP | Comprehensive National Power |
| *CPC | Communist Party of China |
| DA | Department of the Army |
| DOD | Department of Defense |
| ELINT | electronic intelligence |
| *ERA | explosive-reactive armor |
| EW | electronic warfare |
| FM | field manual |
| HE | high explosives |
| HUMINT | human intelligence |
| IADS | integrated air defense system |
| ICBM | intercontinental ballistic missile |
| *IFV | infantry fighting vehicle |
| IO | information operations |
| IW | information warfare |
| JP | joint publication |
| KDP | key defense point |

| | |
|---|---|
| km | kilometer |
| *LC | local command |
| LOC | line of communications |
| m | meter |
| MANPADS | man-portable air defense system |
| *MBT | main battle tank |
| *MD | military district |
| mm | millimeter |
| *MPS | Ministry of Public Security |
| MRL | multiple rocket launcher |
| *MSS | Ministry of State Security |
| NATO | North Atlantic Treaty Organization |
| NCO | noncommissioned officer |
| *PAP | People's Armed Police |
| *PLA | People's Liberation Army |
| *PLAA | People's Liberation Army Army |
| *PLAAF | People's Liberation Army Air Force |
| *PLAJLSF | People's Liberation Army Joint Logistics Support Force |
| *PLAN | People's Liberation Army Navy |
| *PLANMC | People's Liberation Army Navy Marine Corps |
| *PLARF | People's Liberation Army Rocket Force |
| *PLASSF | People's Liberation Army Strategic Support Force |
| PMESII-PT | political, military, economic, social, information, infrastructure, physical environment, and time |
| *PRC | People's Republic of China |
| SAM | surface-to-air missile |
| SHORAD | short-range air defense |
| SIGINT | signals intelligence |
| SOF | special operations forces |
| *SPAAG | self-propelled antiaircraft gun |
| *SPG | self-propelled gun |
| SRBM | short-range ballistic missile |
| *TC | theater command |
| TEL | transporter-erector-launcher |
| UA | unmanned aircraft |
| UAS | unmanned aircraft system |
| U.S. | United States |
| USAF | United States Air Force |
| USSOCOM | United States Special Operations Command |

## SECTION II – TERMS

**course of action**

A scheme developed to accomplish a mission. (JP 5 0)

This page intentionally left blank.

# References

All websites accessed on 11 May 2021.

## REQUIRED PUBLICATIONS

These documents must be available to the intended user of this publication.

*DOD Dictionary of Military and Associated Terms.* January 2021.

FM 1-02.1. *Operational Terms.* 09 March 2021.

FM 1-02.2. *Military Symbols.* 10 November 2020.

## RELATED PUBLICATIONS

These sources contain relevant supplemental information.

### STRATEGIC GUIDANCE AND POLICY

*The Capstone Concept for Joint Operations: Joint Force 2030. Classified and available on SIPRNET only.*

*The National Security Strategy of the United States. 18 December 2017.* Available at https://trumpwhitehouse.archives.gov/wp-content/uploads/2017/12/NSS-Final-12-18-2017-0905.pdf.

### JOINT PUBLICATIONS

Most joint doctrinal publications are available online at the Joint Electronic Library (JEL) at https://www.jcs.mil/Doctrine/Joint-Doctrine-Pubs/.

JP 3-0. *Joint Operations.* 17 January 2017.

JP 5-0. *Joint Planning.* 01 December 2020.

### ARMY PUBLICATION

Army doctrine and training publications are available at https://armypubs.army.mil/.

ADP 3-0. *Operations.* 31 July 2019.

AR 350-2. *Operational Environment and Opposing Force Program.* 19 May 2015.

FM 3-0. *Operations.* 06 October 2017.

FM 6-27/MCTP 11-10C. *The Commander's Handbook on the Law of Land Warfare.* 7 August 2019.

### AIR FORCE PUBLICATION

*Air Force Future Operating Concept: A View of the Air Force in 2035.* September 2015. Available at https://www.af.mil/Portals/1/images/airpower/AFFOC.pdf.

### OTHER PUBLICATIONS

The Army Training Department of the Headquarters of the [Chinese] General Staff. *Infantry Unit Tactics*, 2nd ed. Translated by the National Ground Intelligence Center. 2004.

Sun Tzu. *The Art of War.* Translated by Samuel B. Griffith. London: Oxford University Press, 1963.

## RECOMMENDED READINGS

Blasko, Dennis J. *The Chinese Army Today: Tradition and Transformation for the 21st Century*, 2nd ed. New York: Routledge, 2012.

Ji Rongren, ed. *Services and Arms Application in Joint Operations*, 3rd ed. Translated. Baishan Press, 2010.

The People's Liberation Army. *Army Combined Tactics under the Condition of Informationization*, 2nd ed. Translated by the National Ground Intelligence Center. Shijiazhuang Army Command Academy Press, 2009.

U.S. Defense Intelligence Agency. *China Military Power: Modernizing a Force to Fight and Win*. 15 January 2019. Available at https://www.dia.mil/Military-Power-Publications.

Zhang Yuliang, chief ed. *The Science of Campaigns*. Translated. Beijing, China: National Defense University Press, 2006.

## PRESCRIBED FORMS

This section contains no entries.

## REFERENCED FORMS

Unless otherwise indicated, DA forms are available on the Army Publishing Directorate website at https://armypubs.army.mil.

DA Form 2028. *Recommended Changes to Publications and Blank Forms.*

# Index

Entries are by paragraph number.

**Entries are by paragraph number.**

**Entries are by paragraph number.**

psychological attack, 5-19

information defense, electromagnetic protection, 5-21

information warfare tactics, 5-14–5-24

intelligence protection, 5-24
network protection, 5-22
physical protection, 5-23

information operations characteristics, high target value, 5-5

integration and synthesis importance, 5-6
linkage between attack and defense, 5-7
tactical information operations, 5-3–5-7
universal permeation, 5-4

information operations overview, 5-1–5-2

information operations principles, achieve synthesis, 5-11–5-12
actively attack, 5-9
protect tightly, 5-13
tactical information operations, principles, 5-8–5-13
target nodes, 5-10

information warfare tactics, information attack, 5-15–5-19

information defense, 5-20–5-24
tactics of information operations, 5-14–5-24

informationized battlefield and antiterrorism operations, antiterrorism and stability actions, 9-2–9-7
asymmetry, 9-7
complexity, 9-4
joint nature, 9-6
significant effects, 9-5
urgency, 9-3

informationized battlefield and defensive operations, defensive actions, 8-2–8-6
fewer traditional advantages, 8-4

increasing arduousness, 8-3

increasing importance of offensive actions, 8-6
more dynamic, 8-5

informationized battlefield and offensive operations, cost of modern war, 7-10

electronic warfare importance, 7-7

informationized battlefield transparency, 7-5
multidimensional battlefield, 7-9
offensive actions, 7-4–7-10
precision munitions proliferation, 7-6
rapid tempo of operations, 7-8

informationized battlefield and stability and security actions, antiterrorism and stability actions, 9-29–9-35
hybrid warfare, 9-33
international peacekeeping, 9-35
political nature, 9-32
threat strength, 9-34

informationized battlefield transparency, 7-5

initiate, 7-35–7-38

integrate psychological attack and protection, 5-35

integration, 8-10 integration and synthesis

importance, 5-6

intelligence protection, 5-24

international peacekeeping, 9-35

issue orders, 4-11

**J**

joint actions, 9-11

joint capabilities, 3-1–3-28
air force, 3-1–3-9
navy, 3-10–3-17
rocket force, 3-18–3-23
strategic support force, 3-24–3-28

joint nature, 9-6

**K**

keep defensive zone small, 8-49

key-point concentration, 9-38

**L**

legality, 9-12

light armored and wheeled vehicles, A-27–A-28

light helicopters, D-9 linear search, 6-42

linkage between attack and defense, 5-7

long duration, 5-32

long-range missile systems, C-17–C-19

long-range raid, 7-117

**M**

main force security, 9-55–9-56
maintain concealment and secrecy, 5-38

maintain contact with the opponent, 6-53

maintain freedom of navigation, 1-24

maintain internal security and stability, 1-21

maintain regional stability, 1-23

maintain social order, 9-50

make decisions, 4-10

maneuver capabilities, A-1–A-38
capabilities and limitations, A-34–A-38
equipment overview, A-3–A-33
functional overview, A-1–A-2

man-portable missile systems, C-9–C-12

man-portable missiles and rockets, A-32–A-33

Mao Zedong, 1-30, 1-35–1-36, 1-39, 1-43, 1-51, 1-53, 1-68, 4-1, 7-2, 7-12, 7-82

medium-range missile systems, C-15–C-16

mission-supporting, 6-15

mobile artillery group, 4-55
mobile defensive operations, assume command and communication challenges, 8-58
assume large defensive zone, 8-55
defensive operations types, 8-46–8-127
emphasize flexibility and mobility, 8-57
focus on the offense, 8-56
plan and execute quickly and decisively, 8-54

more dynamic, 8-5 mortars, B-6–B-7

multidimensional battlefield, 7-9

multiple rocket launchers, B-14–B-15

**Entries are by paragraph number.**

**Entries are by paragraph number.**

Made in the USA
Las Vegas, NV
17 December 2023